THERMAL RADIATION

METALS, SEMICONDUCTORS, CERAMICS, PARTLY TRANSPARENT BODIES, AND FILMS

TEMPERATURNOE IZLUCHENIE METALLOV I NEKOTORYKH VESHCHESTV

ТЕМПЕРАТУРНОЕ ИЗЛУЧЕНИЕ МЕТАЛЛОВ И НЕКОТОРЫХ ВЕШЕСТВ

THERMAL RADIATION
Metals, Semiconductors, Ceramics, Partly Transparent Bodies, and Films

Darii Yakovlevich Svet

Authorized translation from the Russian

Springer Science+Business Media, LLC

ISBN 978-1-4899-4961-5   ISBN 978-1-4899-4959-2 (eBook)
DOI 10.1007/978-1-4899-4959-2

The Russian text was published by Metallurgiya Press in Moscow in 1964.

Library of Congress Catalog Card Number 65-25260

CONTENTS

CONTENTS

PREFACE TO THE AMERICAN EDITION

For metals and many other substances existing in the solid and liquid phases, the spectral distribution of the density of thermal (temperature) radiation and its integral are usually characterized quantitatively by the coefficients of spectral and integral radiating power. The values of these coefficients determine the amount of energy involved in radiant transfer, which plays an important part in high-temperature processes of metallurgy, modern chemistry, and reaction techniques. A knowledge of the radiating powers is very necessary in pyrometry, which constitutes the basis of automatic temperature control and other radiation measurements. In the Rayleigh-Jeans, microwave part of the spectrum, the radiating-power coefficients may often by obtained by calculation from classical electrodynamics.

Starting from the near and sometimes the middle infrared part of the spectrum, owing to the phenomenon of anomalous dispersion, for the majority of real substances (in particular metals) these calculations require a considerable development of the apparatus of quantum electrodynamics, and are not very feasible in practice; hence, we shall in general be considering their experimental determination.

Fairly recently, the author has become acquainted with the results of such experimental studies carried out in various countries and scattered through many periodicals.

In preparing the present book, which above all is of a reference-book nature, an attempt has been made to systematize the experimental data, setting them out in a convenient form suitable for use by a wide circle of interested specialists.

The extensive existing literature on the physics of radiation and the theory of interaction between electromagnetic radiation and the electrons of matter (especially metal) meant that it was unnecessary for the author to trouble with theoretical questions. Notes of a theoretical nature as well as certain formulas have been included in order to clarify the physical essence of various data and their functional dependence.

The information on the radiating and reflecting powers has been taken mainly from recent work. In some cases, however, the absence of new data has forced us to turn to older material (Ornstein, Worthing, and a few other workers). The book also contains a brief account of some extremely interesting but insufficiently known theoretical work of the reflection of electromagnetic waves by a rough surface (Brekhovskikh) and the radiation of a partly transparent layer (McMahon).

The description of apparatus and details of measuring methods were not a subject for the present book; the bibliography at the end, although detailed, is not exhaustive.

PREFACE TO THE AMERICAN EDITION

CHAPTER I

RADIATING AND REFLECTING POWER

Any real physical body differs from an absolute black body in that it not only absorbs (emits) but also reflects, and in general also transmits, electromagnetic radiation. In view of this, the thermal radiation of real bodies always differs from that of a black body. According to Kirchhoff's law, the thermal radiation of non-black bodies is also completely determined if their radiating (absorbing) power is known.

For real nontransparent bodies, including metals, the spectral distribution of the energy of radiation with given wavelength λ at temperature T may be described as the produce of the Planck radiant-energy spectral-distribution function $b_0(\lambda, T)$ and some function $\varepsilon(\lambda, T)$ characterizing the spectral monochromatic (radiating) power of the surface in specific assigned conditions.*

Denoting the spectral energy-distribution function of the radiation of an actual body by $b(\lambda_i, T)$ and that of an absolute black body by $b_0(\lambda_i, T)$, the spectral-distribution function of the monochromatic radiating power $\varepsilon(\lambda_i, T)$ may be given as

$$\varepsilon(\lambda_i, T) = \frac{b(\lambda_i, T)}{b_0(\lambda_i, T)}.$$

The radiating power is a measure of the amount of radiant energy emitted by a given surface of some material as compared with the radiant energy emitted by an absolute black body at the same temperature. For a non-transparent body, transmission coefficient $\tau(\lambda, T) = 0$, and hence, from Kirchhoff's law

$$\varepsilon(\lambda_i, T) + \rho(\lambda_i, T) = 1,$$
$$\varepsilon(\lambda_i, T) = 1 - \rho(\lambda_i, T),$$

where $\rho(\lambda_i, T)$ is the spectral-distribution function of the reflection coefficient. The radiating power characterizes a given surface, and not the substance (material) itself. For a nontransparent body, the smallest value of radiating power is characterized by the reflection coefficient of the substance with an ideal mirror surface.

Radiation and Reflection of a Rough Surface

An analysis of the reflection of a plane wave from a rough surface having the form of a lattice was first carried out by Rayleigh [2]. The results obtained showed that, for a degree of roughness much smaller than the wavelength, the regularly-reflected wave of intensity and phase corresponding to ideal mirror reflection was accompanied by diffracted rays. The extent and direction of these may be determined quantitatively as a function of the intensity, incident angle, and wavelength of the incident wave.

As a result of irregular thermal motion (thermal fluctuations), the surface of a liquid is also continuously deformed. Thus diffuse reflection is always present as well as mirror reflection from the surface of a liquid. According to Rayleigh's theory, the presence of roughnesses very small compared with the wavelength is sufficient to produce considerable diffuse reflection.

The question of the rough surface of a liquid was considered by L. I. Mandel'shtam [3].

---

*The reader may acquaint himself with the fundaments of thermal radiation in one of the Optics courses, for example, that of Landsberg [1].

Studying the diffraction of waves at an uneven surface, L. M. Brekhovskikh [4] obtained results for the extremely important practical cases of reflection by a surface of finite dimensions. In this analysis, constituting in essence a generalization of Rayleigh's method, the unevennesses were ascribed sinusoidal form.

Figures 1-8 present polar diagrams of the diffracted wave for two incident angles of the incident light (45 and 80°) and various values of $H = 2\pi a/\lambda$, where a = amplitude of surface "ripples" and λ = wavelength of radiation.

The wavelength L of the "ripples" was taken the same in all cases and equalled 10λ. The width of the area was taken as 4L. We note that the width of the area only affects the width of the lobes. The intensity scale on the graphs is logarithmic (decibels).* This scale is shown beneath each figure. For very small values of a, the ripple is represented by a straight line. The arrows on the diagrams indicate the directions of the incident and reflected waves for mirror reflection.

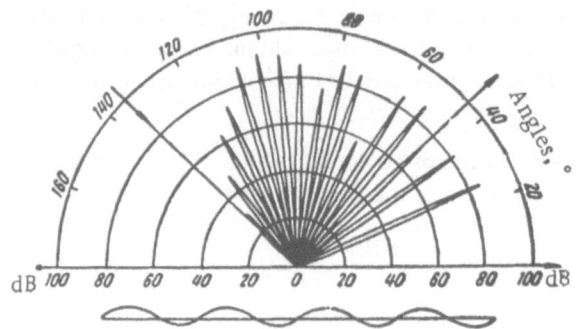

Fig. 1. Rough surface ($2\pi a/\lambda = 6$; L/a = 10.5).

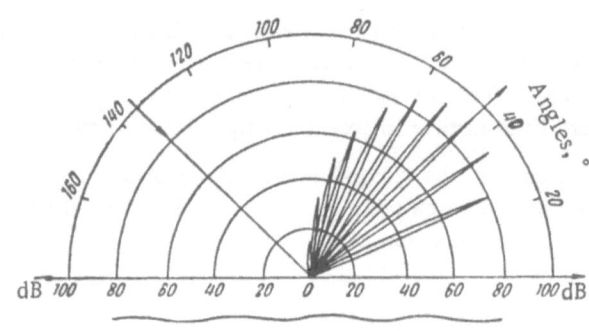

Fig. 2. Rough surface ($2\pi a/\lambda = 1$; L/a = 62.8).

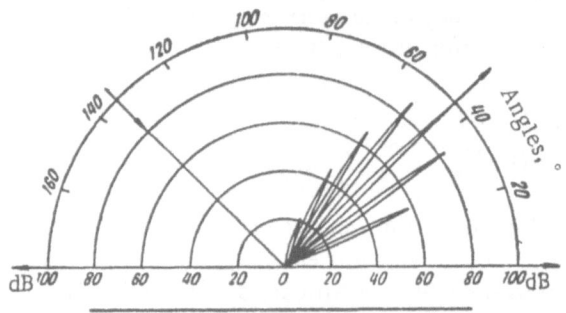

Fig. 3. Rough surface ($2\pi a/\lambda = 0.3$; L/a = 209).

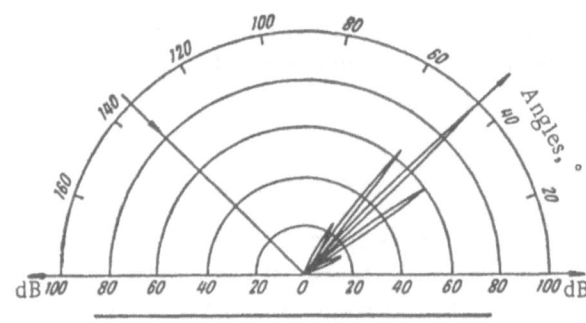

Fig. 4. Rough surface ($2\pi a/\lambda = 0.03$; L/a = 2090).

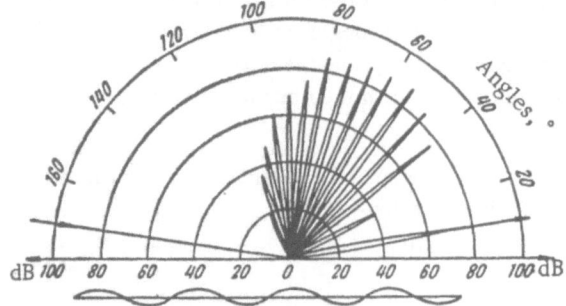

Fig. 5. Rough surface ($2\pi a/\lambda = 6$; L/a = 10.5).

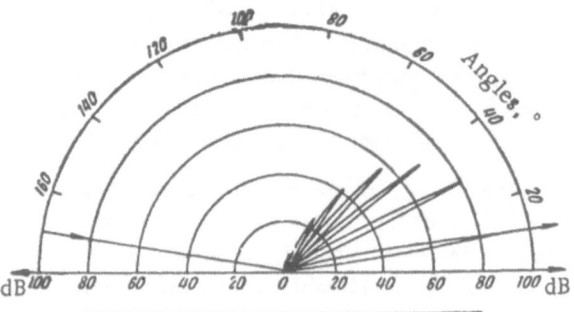

Fig. 6. Rough surface ($2\pi a/\lambda = 1$; L/a = 62.8).

*We remember that transformation into decibels, for example, of some ratio R is effected by means of the formula $R_{dB} = 20 \lg R$.

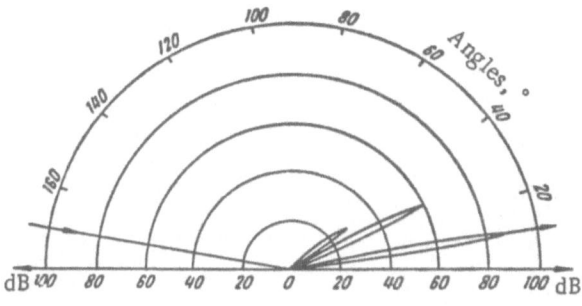
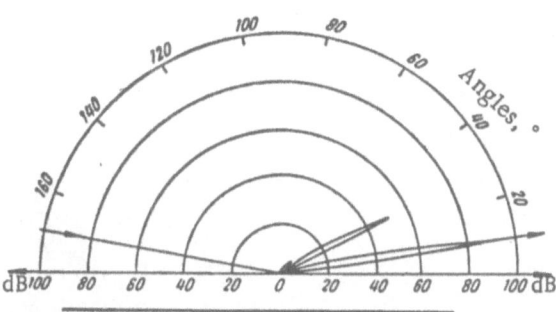

Fig. 7. Rough surface ($2\pi a/\lambda = 0.1$; $L/a = 628$).　　　Fig. 8. Rough surface ($2\pi a/\lambda = 0.03$; $L/a = 2090$).

The results shown indicate that even a slight ripple gives a considerable deviation from mirror reflection (Fig. 3). Reflection can only be considered "mirror" in Fig. 4, where the ripple amplitude is 209 times smaller than the wavelength and 2090 times smaller than L, since here the side lobes are some 60 times (30 dB) smaller than the main one.

In the case of the 80° incident angle, scattering is more directional, for the same values of H, than for the 45° incident angle.

Thus, for small angles of incidence, reflection from a rough surface is more directional.

It should be noted that in actual practice the distribution of unevenness on a rough surface has a random character, so that the figures shown characterize the qualitative rather than the quantitative aspect of the process.*

For every body the spectral energy distribution of the radiation may be presented in the form of curves (isothermal, isochromatic). These curves will differ the more in character from the spectral radiation of a black body, the greater the changes suffered by the function $\varepsilon(\lambda_i, T)$. If the value of function $\varepsilon(\lambda_i, T)$ does not depend on the wavelength, i.e., $\partial\varepsilon(\lambda_i, T)/\partial\lambda = 0$, then the radiation is customarily called "gray."

For gray radiation, the spectral energy-distribution function is similar to the function for a black body.

In actuality, real bodies with gray radiation over the whole range of the spectrum do not exist. Furthermore, over the whole range of the spectrum gray radiation is in principle impossible, but in individual spectral ranges we may with fair accuracy, set $\partial\varepsilon(\lambda_i, T)/\partial\lambda = 0$ for a number of bodies, i.e., we may regard $\varepsilon(\lambda_i, T)$ as constant and the radiation gray. It should be noted that increasing the radiating power of a given substance by changing the geometry of its radiation surface, i.e., reducing the reflection coefficient, for example, on account of the development of roughness, brings the radiation of the surface closer to gray.

Let us suppose that, in a nontransparent body with a mirror-reflecting surface, the reflection coefficients in the λ_1 and λ_2 parts of the spectrum are $\rho(\lambda_1, T) = \rho_1$ and $\rho(\lambda_2, T) = \rho_2$, respectively, and hence, according to Kirchhoff's law, the radiating powers are $\varepsilon_1 = 1 - \rho_1$ and $\varepsilon_2 = 1 - \rho_2$.

Let us assume that, on changing the radiation surface (for example, by corrugation, roughening, etc.), the reflecting power falls m times, i.e., the radiating powers rise and become, respectively, $\varepsilon_1' = 1 - \rho_1/m$ and $\varepsilon_2' = 1 - \rho_2/m$. Clearly, the radiation for the wavelengths chosen (λ_1 and λ_2) will approach more closely to gray, the smaller the difference $\varepsilon_1 - \varepsilon_2 = \Delta\varepsilon$.

In our example, for a smooth surface $\Delta\varepsilon = \varepsilon_1 - \varepsilon_2$ and for a rough $\Delta\varepsilon' = \Delta\varepsilon/m$. But since by hypothesis always m > 1, to $\Delta\varepsilon' < \Delta\varepsilon$, which was to be proved.

On the basis of Kirchhoff's law, the radiating power may be determined if we know what part of the whole incident radiation is absorbed by the substance.

_____

*Reflection for a random surface-roughness distribution was considered by Isakovich [5].

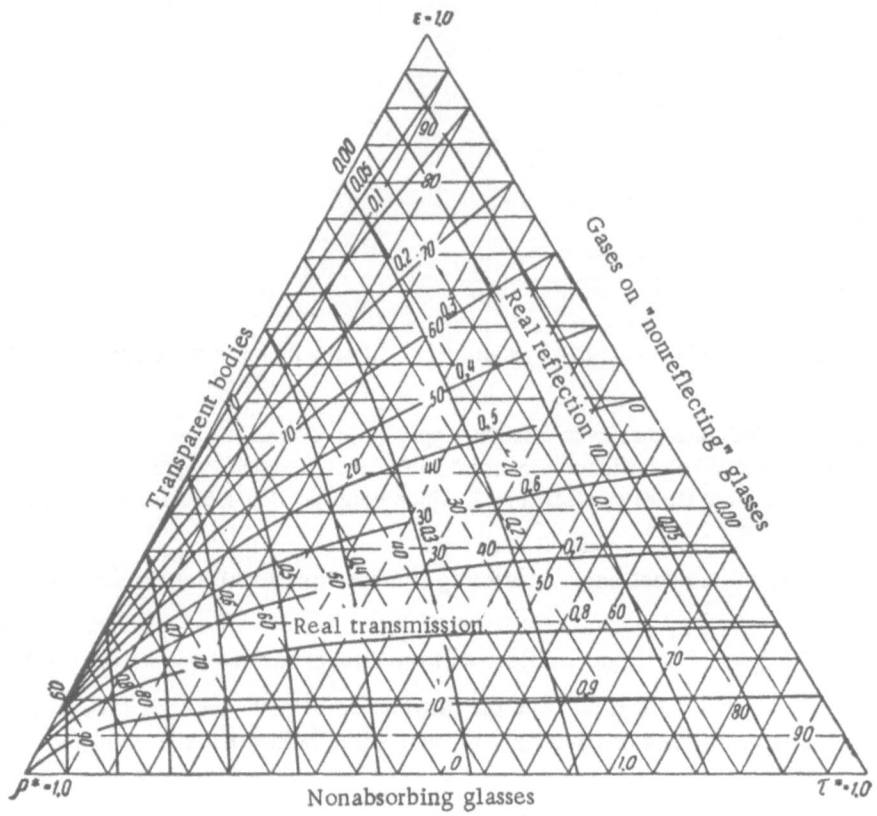

Fig. 9. Diagram of radiating, reflecting, and absorbing powers (McMahon).

Absorption of Radiation in a Layer

Let us consider the absorption of a parallel beam of monochromatic radiation with wavelength λ passing through a homogeneous medium (layer of material) of thickness l. Let us denote the intensity of the incident flux entering the medium by $I_0(\lambda)$. On passing this flux through an infinitely thin plane-parallel layer of matter dx, imagined to be separated out from the thickness of the actual substance, the intensity falls by dI.

Since the elementary layer is chosen inside the substance, losses of radiant energy on account of reflection at the boundary do not occur.

If we treat the relation between the attenuation of radiant energy in the infinitely thin layer and the thickness as linear, we may write dI = $-I(\lambda)\alpha(\lambda)$dx, where $\alpha(\lambda)$ is a factor characterizing the absorption of the layer per unit thickness for given λ, will be the monochromatic-absorption coefficient. Integrating this equation with respect to x between x = 0 and l gives the well-known formula of Lambert: *

$$I(\lambda) = I_0(\lambda) \exp - \alpha(\lambda)\, l.$$

The absorption increases with increasing length. Thus we may write

$$I(\lambda) = I_0(\lambda) \exp \left[-\int_0^x \alpha(\lambda,\, x)\, dx \right].$$

---

*This expression is often given another name [1].

For the waveband, $I(\lambda) = I_0(\lambda)d\lambda$ is the energy-distribution function at the point of incidence of the radiant-energy beam. At a distance x from this point

$$I(\lambda, x)\,d\lambda = I_0(\lambda)\exp - \alpha(\lambda)\,x d\lambda.$$

After integrating with respect to λ, we obtain the total intensity of the radiation which has passed through the layer of thickness x in the wavelength range λ_1 to λ_2

$$I(x) = \int_{\lambda_1}^{\lambda_2} I(\lambda, x)\,d\lambda = \int_{\lambda_1}^{\lambda_2} I_0(\lambda)\exp - \alpha(\lambda)\,x d\lambda.$$

If $I_0(\lambda)$ is the intensity of the total radiation at the point of incidence, then

$$\frac{I(\lambda, x)}{I_0(\lambda, 0)} = \frac{\int_{\lambda_1}^{\lambda_2} I_0(\lambda, 0)\exp - \alpha(\lambda)\,x d\lambda}{\int_{\lambda_1}^{\lambda_2} I_0(\lambda, 0)\,d\lambda}. \qquad (1)$$

If in the waveband λ_1, λ_2 the value $\partial\alpha(\lambda)/\partial\lambda = 0$ and the energy-distribution function does not vary along the beam, equation (1) confirms Lambert's law. In many practical cases $\alpha(\lambda)$ changes very little over the waveband, and Lambert's law holds to a fair approximation. Absorption in substances with a bright-line spectrum deviates considerably from Lambert's law over a wide ($\lambda_2 - \lambda_1$) band. For a waveband narrow in comparison with the breadth of the spectral lines, Lambert's law may be applicable, for even between the lines there is a continuous absorption background on account of their finite width and overlapping "tails." If, however, several absorption lines are included in the observed range $\lambda_2 - \lambda_1$, Lambert's law will be inapplicable.

Effect of Radiation Scattering

If a body is inhomogeneous, then scattering is added to pure absorption. On condition that the dimensions of the scattering particles are small compared with the wavelength, Rayleigh's law is obeyed. According to this law, the scattering intensity is proportional to λ^{-4}. On increasing the size of the particles, the scattering intensity changes more slowly as a function of wavelength.

In order to determine the radiation energy loss due to scattering, we use the value of the scattering cross section Q_S for the scattering center.

This quantity is defined in such a way that the energy collected by an area Q_S normal to the flow of radiation corresponds to the energy scattered by the particle.

Thus for N particles per unit volume the partial loss of energy from the flux in an infinitely short distance Δx will be $NQ_S\Delta x$. Here it is assumed that multiple scattering may be neglected.

For spherical particles the value of Q was calculated by Mie [6]. If the radius of the spherical scattering particle is $a \ll \lambda$, then Rayleigh's formula is valid. If the substance of the scattering particles (of spherical form is a dielectric with dielectric constant ε_2, while the medium is a substance with a dielectric constant ε_1, then, for $a \ll \lambda_1$

$$Q_S = \frac{2^7\pi^5 a^6}{3\lambda_1^4}\left(\frac{\varepsilon_2 - \varepsilon_1}{\varepsilon_2 + 2\varepsilon_1}\right),$$

where λ_1 is the length of the electromagnetic wave in the medium of dielectric constant ε_1.

If the scattering particles are conducting spheres, then, for $a \ll \lambda$

$$Q_S = \frac{5 \cdot 2^5 \pi^5 a^6}{3\lambda^4} .$$

If the radius of the sphere is comparable with the wavelength, then the quantity Q_S oscillates with the rising λ around a mean value of the same order as the area of the projection of the sphere.

For bodies of nonspherical form, the linear dimensions of which are smaller than the wavelength, it is impossible to calculate an exact value for Q_S, but this will be close to the value of the scattering cross section of the equivalent sphere.

Radiation of Partly Transparent Bodies

For bodies which are partly transparent owing to good transmission characteristics or small sample thickness, Kirchhoff's law in its general form is inapplicable. This follows from the fact that Kirchhoff's law is derived on the assumption of thermal equilibrium between the energies of emission and absorption for equal temperatures. For a nontransparent body, the incident energy will be reflected or absorbed and partly emitted. For partly transparent bodies, part of the incident energy will be simply transmitted by the body, which breaks the thermal equilibrium of the system. The transmission, which we may define as the ratio of the radiant energy transmitted by the body to the whole energy falling upon it, is one of the main characteristics of partly transparent bodies, and must be considered in an expression more general than the above law of Kirchhoff, which must represent a particular case of this general expression for zero transmission. The basic principles of the theory of the partially transparent body, which we shall consider later, were developed by McMahon [7].

Energy is radiated and absorbed simultaneously. The radiation generated inside a transparent body must reach its surface. In passing through the thickness of the material, the radiation is attenuated, being absorbed by the material in accordance with Lambert's law. Then part of the radiation energy is reflected by the surface inside the body (reflected back).

In a transparent plate with plane parallel boundaries, the radiation generated from 1 cm^2 of material dx cm thick corresponds to $I(\lambda, T)$dx, where $I(\lambda, T)$ is the total spectral power of the radiation, i.e., the amount of energy radiated in 1 sec from 1 cm^2 in the wavelength range λ to $(\lambda_1 + d\lambda)$.

For unit solid angle, the intensity of the radiation is $0.25\pi I(\lambda, T)$ dx, and for an infinitely small solid angle $d\omega$, for which the radiation is normal to the surface of the plate, the intensity of the radiation is $0.25\pi I(\lambda, T)$ dx $d\omega$.

If the hypothetical (radiating) layer lies at a distance x from the surface of the body, then a quantity of radiant energy diminished by $e^{-\alpha(\lambda, T)x}$ dx $d\omega$, where $\alpha(\lambda, T)$ is a function of the values of absorption coefficient for given wavelength and temperature. Then the intensity of the radiation reaching the surface from this layer will be

$$0.25\pi I(\lambda, T) e^{-\alpha(\lambda, T)x} dx d\omega.$$

The total radiation falling normal to the surface is found by summing this expression for all layers from x = 0 to x = d, where d is the thickness of the plate.

Thus the total normal radiation will be

$$\frac{1}{4\pi} I(\lambda, T) d\omega \int_0^d e^{-\alpha(\lambda, T)x} dx$$

or after integration

$$\frac{I(\lambda, T)}{4\pi\alpha(\lambda, T)} \left[1 - e^{-\alpha(\lambda, T)d}\right] d\omega.$$

The quantity $\exp - \alpha(\lambda, T)$ d determines the transmission of the plate. Denoting this transmission by $\tau(\lambda, T)$, we may rewrite the preceding expression in the form

$$\frac{I(\lambda, T)}{4\pi\alpha(\lambda, T)} \left[1 - \tau(\lambda, T)\right] d\omega.$$

Then the radiation which penetrates through the surface with reflecting power $\rho(\lambda, T)$ and is emitted by this surface will be

$$[1 - \rho(\lambda, T)] \frac{I(\lambda, T)}{4\pi\alpha(\lambda, T)} [1 - \tau(\lambda, T)] d\omega.$$

If the body is nontransparent owing to a large value of $\alpha(\lambda, T)$ or d, then Kirchhoff's law is applicable in general form (see p. 3).   For this case $\tau(\lambda, T) = 0$, and hence we may write

$$[1 - \rho(\lambda, T)] \frac{I(\lambda, T)}{4\pi\alpha(\lambda, T)} d\omega = \frac{I_0(\lambda, T)}{2\pi} [1 - \rho(\lambda, T)] d\omega,$$

where $I_0(\lambda, T)$ is the spectral radiation power of an absolutely black body. Hence we may conclude that for $\tau(\lambda, T) = 0$

$$\frac{I(\lambda, T)}{\alpha(\lambda, T)} = 2I_0(\lambda, T).$$

This expression shows that the ratio of the total emissive power to the absorption coefficient equals twice the radiation power of a black body for the same wavelengths and temperatures. The factor of 2 appears because in determining the strength of the black-body radiation we considered the radiation propagating in the direction of a solid angle of 2π steradians, while for the power radiated by the volume this angle equals 4π steradians.

Returning to the expression for the total nonreflected part of the radiation which is created by the lower layers and penetrates through the surface of the transmitting plate, we may write

$$[1 - \rho(\lambda, T)] \frac{I_0(\lambda, T)}{2\pi} [1 - \tau(\lambda, T)] d\omega.$$

The reflected part of the radiation passes through the layer of material d thick, then is again reflected from the far side of the layer and again passes through the thickness of d cm.

The part of this energy coming out of the layer for the first time again suffers the distribution indicated above.

The intensity of this part of the radiation after passing twice through the layer d equals

$$\rho^2(\lambda, T)\tau^2(\lambda, T) [1 - \rho(\lambda, T)] \frac{I_0(\lambda, T)}{2\pi} [1 - \tau(\lambda, T)] d\omega.$$

Then the reflection process is repeated again, and for the third passage through the layer we shall have

$$\rho^4(\lambda, T) \tau^4(\lambda, T) [1 - \rho(\lambda, T)] \frac{I_0(\lambda, T)}{2\pi} [1 - \tau(\lambda, T)] \, d\omega.$$

The total energy which has come out of all the layers amounts to

$$\sum_{n=2p}^{n=\infty} \rho^n(\lambda, T) \tau^n(\lambda, T) \frac{I_0(\lambda, T)}{2\pi} [1 - \tau(\lambda, T)] \, d\omega,$$

where p = 0, 1, 2, 3,..., ∞.

Here we have considered only the radiation emitted in the direction of the forward surface. The radiation emitted toward the back surface is also partly reflected by this. This reflected part of the energy is also emitted through the forward surface. It may be shown analogously that the total radiation of all the components to the back surface may be written in the form

$$\sum_{n=2p+1}^{n=\infty} \rho^n(\lambda, T) \tau^n(\lambda, T) [1 - \rho(\lambda, T)] \frac{I_0(\lambda, T)}{2\pi} [1 - \tau(\lambda, T)] \, d\omega,$$

where p = 0, 1, 2, 3,..., ∞.

The total radiation of the forward surface may be obtained if we add both sums with even and odd values of n. Of course; this resultant sum equals the spectral power of the radiation in the direction of the solid angle dω, i.e.,

$$I(\lambda, T) \frac{d\omega}{2\pi} = \sum_{n=0}^{n=\infty} \rho^n(\lambda, T) \tau^n(\lambda, T) [1 - \rho(\lambda, T)] \frac{I_0(\lambda, T)}{2\pi} [1 - \tau(\lambda, T)] \, d\omega,$$

where n = 0, 1, 2,..., 2p, 2p + 1.

Considering that $\rho(\lambda, T)$ and $\tau(\lambda, T)$ are always smaller than unity, the last expression may be put in the following form:

$$I(\lambda, T) = \frac{[1 - \rho(\lambda, T)] [1 - \tau(\lambda, T)] I_0(\lambda, T)}{[1 - \rho(\lambda, T) \tau(\lambda, T)]}.$$

The radiating power is also defined by the value of the ratio

$$\varepsilon(\lambda, T) = \frac{b(\lambda, T)}{b_0(\lambda, T)} = \frac{I(\lambda, T)}{I_0(\lambda, T)}.$$

Thus we may write

$$\varepsilon(\lambda, T) = \frac{[1 - \rho(\lambda, T)] [1 - \tau(\lambda, T)]}{[1 - \rho(\lambda, T) \tau(\lambda, T)]}.$$

If we put $\tau(\lambda, T) = 0$, which corresponds to the radiation of a nontransparent body, when

$$\varepsilon(\lambda, T) = 1 - \rho(\lambda, T),$$

10

i.e, as we should expect, we obtain the expression for Kirchhoff's law in general form. If, however, $\rho(\lambda, T)$ (in the case of "nonreflecting glasses" or incandescent gases), then in the first approximation $\varepsilon(\lambda, T) = 1 - \tau(\lambda, T)$. This result is in complete agreement, for example, with the conclusions of Finkelburg [8].

The normal reflecting power of unit surface of a transparent plate will be $\rho(\lambda, T)$. The apparent reflecting power $\rho*(\lambda, T)$, however, is always larger than the normal, owing to multiple internal reflections, which increase the fundamental value of the reflecting power.

If the intensity of the normally incident flux is $I(\lambda, T)$, then the intensity of the main reflected flux will be $I(1, T)\rho(\lambda, T)$. Part of the incident flux, a quantity $[1 - \rho(\lambda, T)]I(\lambda, T)$, penetrates through the surface and falls to the value $I(\lambda, T) [1 - \rho(\lambda, T)]\tau(\lambda, T)$ after passing through the body. This flux, having passed through the surface of the body, it partly reflected back with an intensity $I(\lambda,T)[1 - \rho(\lambda, T)]\tau(\lambda, T)\rho(\lambda, T)$; it falls on the side opposite to the forward surface, and is reduced by an amount $[1 - \rho(\lambda, T)]\tau^2(\lambda, T)\rho^2(\lambda, T)$ as compared with the original intensity $I(\lambda, T)$.

Thus the penetrating intensity of the radiation, as a result of which the main reflecting power is increased, will be $I(\lambda, T)[1 - \rho(\lambda, T)]\tau^2(\lambda, T)\rho^2(\lambda, T)$. After this, part of the energy is again reflected from the surface of the body; this enabled McMahon to describe the resulting values of fluxes in the form of series. The over-all sum of these series gives the total intensity of the reflected energy

$$I(\lambda, T)\rho*(\lambda, T) = I(\lambda, T)\rho(\lambda, T) + \sum_{n=1}^{n=\infty} \frac{I(\lambda, T)[1 - \rho(\lambda, T)]^2}{\rho(\lambda, T)} [\rho^2(\lambda, T)\tau^2(\lambda, T)]^n.$$

The exact value of the right-hand side of this expression will be

$$I(\lambda, T)\rho(\lambda, T) + \frac{I(\lambda, T)[1 - \rho(\lambda, T)]^2\rho(\lambda, T)\tau^2(\lambda, T)}{1 - \rho^2(\lambda, T)\tau^2(\lambda, T)}.$$

The apparent reflecting power $\rho*(\lambda, T)$ is obtained from this quantity on dividing by $I(\lambda, T)$, i.e.,

$$\rho*(\lambda, T) = \rho(\lambda, T)\left\{1 + \frac{\tau^2(\lambda, T)[1 - \rho(\lambda, T)]^2}{1 - \rho^2(\lambda, T)\tau^2(\lambda, T)}\right\}.$$

It should be noted that for a perfectly transparent body, when $\tau(\lambda, T) = 1$, the expression for $\rho*(\lambda, T)$ takes the form

$$\rho*(\lambda, T) = \frac{2\rho(\lambda, T)}{1 + \rho(\lambda, T)}.$$

This expression shows that the apparent reflecting power of a transparent body approximately equals twice the true reflecting power of the body, which in this case is extremely small in absolute magnitude, and only for $\tau(\lambda, T) = 0$ are the apparent and true reflecting powers identical.

For a transparent plate with reflecting surfaces, the apparent transmission is in turn smaller than the true value, owing to reflection losses. If the intensity of the incident beam is $I(\lambda, T)$, then it falls to $I(\lambda, T)[1 - \rho(\lambda, T)]$ as a result of reflection.

Thus the energy coming out of the opposite side of the plate changes its intensity as compared with the initial value by $[1 - \rho(\lambda, T)]^2\tau(\lambda, T)$ times; in the same way as before, we may find an expression for the energy allowing for multiple reflections. This expression will have the form:

$$I(\lambda, T)[1 - \rho(\lambda, T)]\tau^2(\lambda, T)\sum_{n=0}^{n=\infty} \rho^{2n}(\lambda, T)\tau^{2n}(\lambda, T).$$

Considering that always $\rho(\lambda, T) < 1$ and $\tau(\lambda, T) < 1$ the expression for the apparent transmitting power $\tau^*(\lambda, T)$ may, after the necessary transformations, be written in the form

$$\tau^*(\lambda, T) = \tau(\lambda, T) \frac{[1 - \rho(\lambda, T)]^2}{1 - \rho^2(\lambda, T)\, \tau^2(\lambda, T)} \; .$$

Ordinary measurements of transmitting power, defined as the ratio of the emergent and incident fluxes, will contain an error if the result obtained is interpreted as $\tau(\lambda, T)$, as is often done in radiation pyrometry. The reflecting power of transparent materials in the high-absorption range, however, is usually extremely small, so that the error will not be very great.

Thus, according to McMahon, the quantities in general characterizing the radiation of any body are:

The radiating power

$$\varepsilon(\lambda, T) = \frac{[1 - \rho(\lambda, T)][1 - \tau(\lambda, T)]}{[1 - \rho(\lambda, T)\, \tau(\lambda, T)]} \; .$$

The apparent reflecting power

$$\rho^*(\lambda, T) = \rho(\lambda, T) \left\{ 1 + \frac{\tau^2(\lambda, T)[1 - \rho(\lambda, T)]^2}{1 - \rho^2(\lambda, T)\, \tau^2(\lambda, T)} \right\} \; .$$

The apparent transmitting power

$$\tau^*(\lambda, T) = \tau(\lambda, T) \frac{[1 - \rho(\lambda, T)]^2}{[1 - \rho^2(\lambda, T)\, \tau^2(\lambda, T)]} \; .$$

We can show algebraically that the sum of these three expressions always equals unity:

$$\varepsilon(\lambda, T) + \rho^*(\lambda, T) + \tau^*(\lambda, T) = 1 \ldots$$

This conclusion may also be reached on the basis of thermodynamics.

If a plate partly transmitting radiation is in thermal equilibrium, then for all wavelengths the sum of the radiating power, the apparent reflecting power, and the apparent transmitting power will be unity.

The relationship enabled McMahon to construct the nomogram shown in Fig. 9 for determining $\varepsilon(\lambda, T)$, $\rho^*(\lambda, T)$, and $\tau^*(\lambda, T)$ in the form of a triangle. This nomogram contains curves of $\rho(\lambda, T) = $ const (true reflecting power) and $\tau(\lambda, T) = $ const (true transmission), which constitute a solution of the equation for $\rho(\lambda, T)$ and $\tau(\lambda, T)$. The complete characteristics of the radiation of a given body for a given wavelength and definite temperature are represented on this nomogram by a point. The position of the point is determined by any two out of the three quantities $\varepsilon(\lambda, T)$, $\rho^*(\lambda, T)$, and $\tau^*(\lambda, T)$, as reckoned by the numbers placed on the medians. Knowing the position of the point, we can also determine the values of $\rho(\lambda_i, T_i)$ and $\tau(\lambda_i, T_i)$.

For any value of temperature, on changing the wavelength, the reflecting and transmitting powers change also, as a result of which a certain curve will be obtained on the triangular nomogram. Each temperature corresponds to its own curve on this nomogram. The complete radiation properties of the given body are described by two families of intersecting straight lines: One family comprises curves for constant temperature, and the other curves for constant wavelength. It is interesting to note that all the vertices of the triangular nomogram correspond to "ideal" bodies of three different well-known types. The "top" vertex corresponds to $\varepsilon(\lambda, T) = 1$, i.e., characterizes the radiation of an absolutely black body. The "right-hand" vertex characterizes an ideal metallic reflector. Optical vacuum, just as an ideal metal mirror, naturally has zero transmitting power.

Thus on McMahon's nomogram the I axis, $\tau^*(\lambda, T) = 0$, characterizes the radiation of nontransparent bodies, for example, metals, for which Kirchhoff's law is used in its usual form; the II axis, $\rho^*(\lambda, T) = 0$ similarly corresponds to the radiation of gases and nonreflecting glasses; the III axis, $\varepsilon(\lambda, T) = 0$, describes a "non-radiating" body; to this, with a certain tolerance, for definite parts of the spectrum, we may relate certain crystals and glasses.

Spectral and Integral Radiating Power

In radiation calculations, the radiating power is usually considered from two points of view.*

The first point of view requires the determination of the monochromatic radiating power, i.e., the radiating power of the given surface over an extremely narrow spectral range of wavelengths. As already indicated, the monochromatic radiating power may be defined as the ratio between the spectral brightness of the given surface $b(\lambda_i, T)$ and the spectral brightness of a black body $b_0(\lambda_i, T)$ at the same temperature T. Graphically the spectral radiating power may be defined as the ratio of corresponding ordinates of the spectral energy-distribution isotherms of the radiation from the given surface and the black body. In actuality, monochromatic systems do not exist. The radiation fluxes with energies $b(\lambda_i, T)$ and $b_0(\lambda_i, T)$ are in fact quasimonochromatic. The energy of each is enclosed within a spectral range from λ_i' to λ_i''. Thus the expression determining the quasimonochromatic spectral radiating power $\overline{\varepsilon(\lambda_i, T)}$ will be

$$\overline{\varepsilon(\lambda_l, T)} = \frac{\int_{\lambda_l'}^{\lambda_l''} b(\lambda_i, T)\, d\lambda}{\int_{\lambda_l'}^{\lambda_l''} b_0(\lambda_l, T)\, d\lambda} = \frac{\int_{\lambda_l'}^{\lambda_l''} \varepsilon(\lambda_i, T)\, b_0(\lambda_l, T)\, d\lambda}{\int_{\lambda_l'}^{\lambda_l''} b_0(\lambda_i, T)\, d\lambda} . \tag{2}$$

If the radiation is gray in the range $\lambda_i' - \lambda_i''$, i.e.,

$$\left| \frac{\partial \varepsilon(\lambda_i, T)}{\partial \lambda} \right| = 0, \quad \lambda_i' \leqslant \lambda \leqslant \lambda_i''$$

then the value $\varepsilon(\lambda_i, T) = $ const and may be taken outside the integral sign on the right-hand side of expression (2), i.e.,

$$\varepsilon(\lambda_i, T) = \varepsilon(\lambda_E, T),$$

where λ_E is the effective and limiting value of wavelength for the quasimonochromatic system with passband $\lambda_i'' - \lambda_i'$.

From the second point of view, the total thermal effect (per unit surface) of the radiation summed over the whole spectrum from 0 to ∞ is considered. In this case the total radiating power $\varepsilon_\Sigma(T)$ also called the "non-blackness" coefficient) is defined as the ratio of the total energy brightness E_T of the surface of the given body held at constant temperature to the energy brightness E_T° of a black body held at the same temperature; thus

$$\varepsilon_\Sigma(T) = \frac{E_T}{E_T^\circ} .$$

---

*In visual pyrometry G. Ribault [9] also uses the concept of light radiating power determined by comparing the total brightnesses of the given surface and a black body. The color radiating power is a quantity which has also received some attention in pyrometry; it is determined by comparing the radiation from the given surface with that from a black body having the same incandescent color.

In other words, the total radiating power may be defined as the ratio of the area of the spectral brightness-distribution isotherm of the given surface to that of a black body at the same temperature.

Hence,

$$\varepsilon_\Sigma (T) = \frac{\int\limits_0^\infty \varepsilon (\lambda_i, T) \, b_0 (\lambda_i, T) \, d\lambda}{\int\limits_0^\infty b_0 (\lambda_i, T) \, d\lambda} \, , \qquad (3)$$

or, on the strength of the Stefan–Boltzmann law,

$$\int\limits_0^\infty b_0 (\lambda, T) \, d\lambda = \sigma_0 T^4,$$

where $\sigma_0 = 5.775 \cdot 10^{-12}$ W $\cdot$ cm$^{-2} \cdot$ deg^4. Then

$$\varepsilon_\Sigma (T) = \frac{\int\limits_0^\infty \varepsilon (\lambda_i, T) \, b_0 (\lambda_i, T) \, d\lambda}{\sigma_0 T^4} \, . \qquad (4)$$

Clearly for a gray body $\varepsilon(\lambda_i, T) = $ const, and hence $\varepsilon_\Sigma (T_i) = \varepsilon(\lambda_i, T_i)$. In practical pyrometry and thermotechnology, the instrument for determining $\varepsilon_\Sigma(T)$ is very often a "total-radiation" radiation pyrometer. As radiant-energy receiver, to act, in the present case, as "integrator over the spectrum," the radiation pyrometer usually incorporates a thermopile with well-blackened thermo-junctions. Such receivers are nonselective over a wide range of the optical spectrum, i.e., over this range they have constant sensitivity. The spectral characteristic is a straight line parallel to the axis of abscissas. In the majority of modern constructions of radiation pyrometers, however, the optical system uses lenses and not a mirror.

For high temperatures, glass lenses (passband not extended beyond 2.7μ) or quartz lenses ($\lambda_{lim} \leq 4$ μ are universally used.

Naturally in this case the infinite upper limit of integration should not be used in formulas (3) and (4). Instead the upper limit should be λ_{lim}. The radiating power $\varepsilon_\Sigma'(T)$ will thus be a coefficient of partial rather than total radiation, corresponding to the bounded characteristic of the spectral transmission of the glass (or quartz). In a number of papers on the determination of $\varepsilon_\Sigma(T)$ due attention has not been paid to this. The so-determined values of $\varepsilon_\Sigma(T)$, as already indicated, constitute coefficients of partial radiation and lose their universality.* This kind of radiation coefficient characterizes only the radiation of the given body over a limited range of the spectrum. Naturally, the error in this will be the more perceptible, the more selective the character of the radiation from the given body. For very high temperatures (> 3000°C), use of lenses made from ordinary glass restricts the "totality" of the radiation detected on the short-wave side (ultraviolet part of the spectrum). The amount of ultraviolet radiation in the spectrum of a black body does not exceed 1% up to a temperature of 3000°C. It is just in the ultraviolet region, however, that there is a maximum in the spectral variation of $\varepsilon(\lambda, T)$ for a number of substances (including the majority of metals). Lenses of other materials applicable for the objectives of total-radiation pyrometers also have transmission factors varying with the wavelength.

---

*In practice these can be used to pass over to the true temperature only for measurements of "radiation temperature" by total-radiation pyrometers with the same spectral characteristics as those for which $\varepsilon_\Sigma(T)$ was measured.

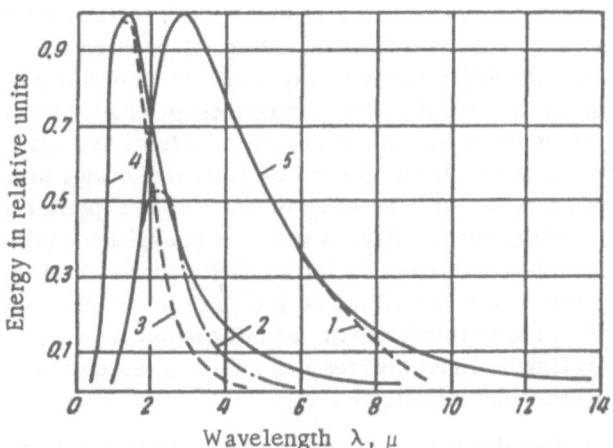

Fig. 10. Radiation energy transmitted by fluorite and glass for a black body at temperatures 1000 and 2000°K. 1) For fluorite at 1000°K; 2) for glass at 1000°K; 3) for glass at 2000°K; 4) black body at 2000°K; 5) black body at 1000°K.

The curves in Fig. 10 constitute the isotherms of a black body for temperature 1000°K after traversing a plate of lithium fluoride (fluorite) 1 cm thick. For this temperature, the ratio of the areas bounded by the transmission curve of the material and the black body isotherm, respectively [total transmission factor $\tau_\Sigma(T)$], corresponds to 0.9. For 2000°K it rises to $\tau_\Sigma(T) = 0.98$.

For optical glass the changes are still more considerable (curves 2 and 3). Thus, for the radiation of a black body at temperature 1000°K the total transmission of glass 1 cm thick is close to 0.3. At a black-body temperature of 2000°K this quantity reaches 0.6. Analogous results are obtained with quartz. It must also be mentioned that, in using a refractive (lens) optical system in radiation pyrometers, an extremely important part is played by chromatic aberration. The focusing of the radiation pyrometer, i.e., bringing the image of the source on to the thermocouple junction, is in fact carried out visually. If, however, the focusing is effected with respect to the visible part of the spectrum, it may not be correct for the infrared. For quartz and glass, on passing from the visible part of the spectrum to the near infrared, the focal length in a number of specific systems of radiation pyrometry changes by 10 mm amd more. Thus, with visual focusing only, some of the rays of the radiation to be measured may not hit the thermojunction.* As a result of chromatic aberrations, the spectral characteristic of any refracting radiation pyrometer has a selective character.

Ignorance of this aspect, often found in practice, may lead to unacceptable errors in using radiation pyrometers for measuring the temperatures of bodies with radiation differing from that of the black body for which the pyrometer was originally calibrated. The limitation of the passband in the optical systems of total-radiation pyrometers in fact makes these pyrometers detectors of partial and not total radiation. For such pyrometers the total radiating power is a quantity characterizing partial and not total radiation:

$$\varepsilon_\Sigma'(T) = \frac{\int_{\lambda_1}^{\lambda_2} \varepsilon(\lambda_i, T)\, b_0(\lambda_i, T)\, d\lambda}{\int_{\lambda_1}^{\lambda_2} b_0(\lambda_i, T)\, d\lambda} . \tag{5}$$

---

*Despite the physically obvious nature of these considerations, irreprovably set forth in the classical monograph of G. Ribault [9], this is frequently forgotten, leading to a number of serious misunderstandings and errors.

Comparing the expression for the total radiating power (5) with expression (2) for the spectral radiating power $\varepsilon(\lambda_i, T)$, we may come to the conclusion that the difference between these is external and bears only a "quantitative character." In fact, the difference lies only in the limits of integration due to the width of the wavelength passband. The band of the partial-radiation brightness pyrometer is always narrower than that of a total-radiation system. This difference in transmission bands often leads to a sharp difference in the values of $\varepsilon_\Sigma(T)$ and $\varepsilon(\lambda_i, T)$ As mentioned earlier, in the general case, for real bodies with broad wave passbands $\lambda_2 - \lambda_1$, the radiation cannot be regarded as gray. Thus for nongray radiation in expression (5) $\varepsilon(\lambda_i, T) \neq$ const. This term can now not be taken outside the integral sign, as was done in expression (2) when determining $\varepsilon(\lambda_i, T)$. A total-radiation system may also be considered as a sort of quasimonochromatic system, ascribing to it a quite definite value of effective limiting wavelength [9]. This gives rise to the question of introducing, for a total-radiation system, the concept of a quasimonochromatic radiating power, to which could be ascribed a definite value of effective limiting wavelength. Naturally, for gray radiation this question does not arise, since $\varepsilon_\Sigma(T) = \varepsilon(\lambda_i, T)$ for every wavelength, including the effective one.

For nongray selective radiation things are different, since the effect of the selectivity of the radiation, described by the function $\varepsilon(\lambda_i, T)$, for a wide transmission band of a total-radiation system, will change the limiting effective wavelength.

In the case of nongray radiation, the wavelength of the quasimonochromatic system corresponding to the center of gravity λ_c, isotherm must be determined with due allowance for the function $\varepsilon(\lambda_i, T)$. In particular, for a total-radiation pyrometer with an infinitely wide passband this must be done from the formula

$$\lambda_c' = \frac{\int_0^\infty \lambda b(\lambda_i, T)\, d\lambda}{\int_0^\infty b(\lambda_i, T)\, d\lambda} = \frac{\int_0^\infty \lambda \varepsilon(\lambda_i, T)\, b_0(\lambda_i, T)\, d\lambda}{\int_0^\infty \varepsilon(\lambda_i, T)\, b_0(\lambda_i, T)\, d\lambda} \ . \tag{6}$$

What we have said regarding the representation of a total-radiation system in the form of a quasimono-chromatic system for nongray radiation may be carried over to a partial-radiation system. In this case, we only need to change the limits of integration in expression (6). Hence the so-called total radiating power character-izes the system only for some part of the spectrum. Hence the total radiating powers given in the text-books (the so-called "nonblackness" coefficients) by no means bear a universal character. It is important to know not only the temperature at which these coefficients were determined by some author or other, but also the method and the passband of the detector, or more precisely the characteristics of its spectral sensitivity. Whereas the nonblackness tables usually give the temperatures, the spectral characteristics of the detector are very often absent, and the published values of $\varepsilon_\Sigma(T)$ must therefore be used with caution.

As an obvious example in which the dependence of the radiating power on the spectral characteristics of the total-radiation detector used for its determination appears in exaggerated form, let us consider the radiation of carbonic-acid gas CO_2. The CO_2 absorption spectrum, which depends on the molecular structure [10], is shown in Fig.11. Clearly the degree of "nonblackness" of CO_2 measured by a radiation pyrometer will vary sharply with the transmission of the material from which the optical part of the telescope is made. For example, the results of measuring $\varepsilon_\Sigma(T)$ by a telescope with a quartz lens are unsuitable for a telescope with a glass lens. In the latter case the whole of the main emission band near $\lambda = 4.2\mu$ will not "go through." Radiation in the region around $\lambda = 2.7\mu$ will also be strongly absorbed by the glass. Use of a mirror telescope shows us a still further (substantial at low temperatures) absorption of CO_2 in the spectral range around 14 to 15μ. Considerable discrepancies may be obtained on using telescopes with different spectral characteristics for measuring the $\varepsilon_\Sigma(T)$ of metals, molten glasses, etc.

The lack of a universal character in the total radiating power $\varepsilon_\Sigma(T)$, so widely used in pyrometry and thermotechnology, justifies us in once more directing the attention of the reader to the fact that, for nongray radiation, i.e., $\partial_\varepsilon(\lambda_i, T)/\partial\lambda \neq 0$, the value of the total radiating power $\varepsilon_\Sigma(T)$, even when taken from tables, may quite inadequately characterize the real radiation process. In fact, for a certain temperature, the radiation is only completely characterized by the actual spectral-distribution function of radiating power $\varepsilon(\lambda_i, T)$.

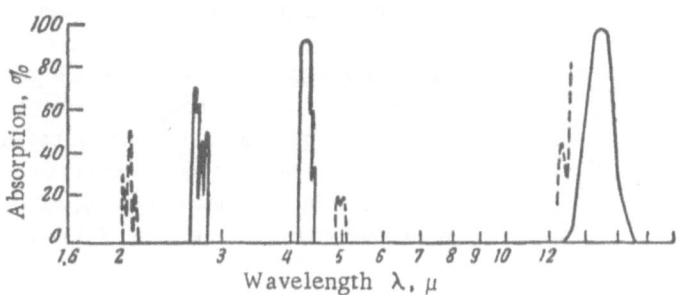

Fig. 11. Absorption spectrum of CO_2 (Barker).

Dependence of Radiating Power on Angle of Radiation

The monochromatic and total radiating powers also depend on the direction of the radiation, characterized by the angle between the direction of the ray and the normal to the radiating surface. According to Lambert's law, the amount of energy transmitted by a diffusely radiating surface in a given direction is proportional to the cosine of the angle (φ) between the ray and the normal. Figure 12 shows a polar diagram characterizing the intensity and density of the radiation emitted by an elementary surface area df. Since the intensity of radiation is the same in all directions, the ends of the intensity vectors lie on a semicircle, or in space on a hemisphere. The radiation density changes on changing the direction, since it is determined by the product of the radiation intensity I and the projection of the area $Idf \cos \varphi$. Hence the ends of the vectors $Idf \cos \varphi$ lie on a circle, or in space on a sphere, tangent to the area $df \cos \varphi$ at its center. It has been established experimentally that Lambert's law is satisfied for diffuse radiation at angles $\varphi \leq 70$.

For large angles the radiation diminishes. The character of this diminution is shown by the broken line in Fig. 12. The radiation of real bodies deviates from Lambert's law. For example, we know from the fundaments of metal optics ([11] or [1]) that the radiation of metals is strongly polarized. We remember that one plane of polarization is determined by the axis of the ray and the normal to the surface, while the second plane of polarization, passing through the same axis, is perpendicular to the first. Depending on the position of the polarization plane, quite different laws for the angular distribution of the radiation may be obtained.

Fig. 12. Lambert's law.

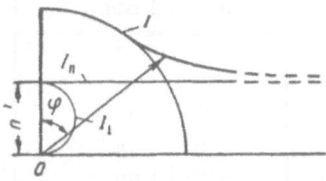

Fig. 13. Distribution of polarized radiation of a metal in polar coordinates: n = refractive index.

A typical form of the curves for the distributions normal and parallel to the radiating plane, $I_\perp$ and $I_\parallel$, respectively, appears in Fig. 13. In this figure the nonpolarized radiation, or so-called natural light, being the sum of the normal and parallel components, is shown by curve I, so that $I = I_\perp + I_\parallel$. Figure 13 also shows that the radiation density from the surface of a polished metal has a minimum in the direction of the normal to the surface and rises on increasing the angle of incidence up to approximately $\varphi = 75°$. For this angle, the radiation density in the parallel direction is roughly double that in the perpendicular. Naturally the radiation density cannot exceed that of a black body. Also dependent on the angle of incidence are the spectral and total radiating powers $\varepsilon(\lambda_i, T)$ and $\varepsilon_\Sigma(T)$. Hence we distinguish two values for each of the two kinds of coefficients (spectral and total): the normal values $\varepsilon_n(\lambda_i, T)$ and $\varepsilon_{\Sigma n}(T)$, and the hemispherical $\varepsilon_c(\lambda_i, T)$ and $\varepsilon_{\Sigma c}(T)$. The normal spectral radiating power $\varepsilon_n(\lambda_i, T)$ may be defined as the ratio of the normal spectral brightness of a polished nontransparent mirror made out of the given material and kept a certain temperature to the normal spectral brightness of a black body held at the same temperature. The

17

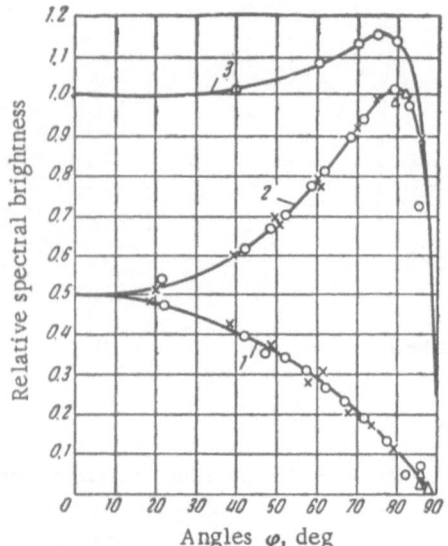

Fig. 14. Relative spectral brightness of tungsten: 1) For polarized light ($I_{\parallel}$); 2) for polarized light ($I_{\perp}$); 3) for natural light ($I_{\perp} + I_{\parallel}$).

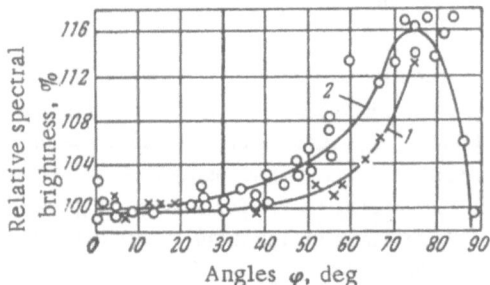

Fig. 15. Variation of spectral brightness. 1) For wavelength $\lambda = 0.47\ \mu$; 2) for wavelength $\lambda = 0.66\ \mu$.

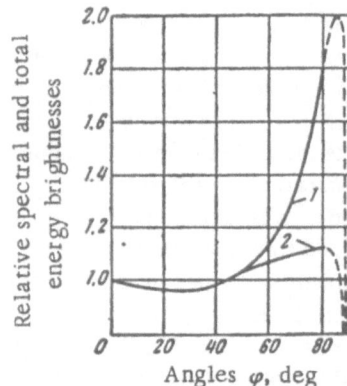

Fig. 16. Variation of total and spectral (red part of spectrum) radiating power of platinum with emission angle: 1) Total; 2) spectral.

spectral hemispherical radiating power $\varepsilon_c(\lambda_i, T)$ may be defined as the ratio of the luminance of a polished mirror made of the given substance and kept at a given temperature to the luminance of a black body at the same temperature. Having the value of the spectral radiating power $\varepsilon_\varphi(\lambda_i, T)$ for various angles φ, and noting that the projection of the element of area varies in proportion to $\cos\varphi$, we may calculate the hemispherical radiating power from the expression

$$\varepsilon_c(\lambda_i, T_i) = \frac{\int_0^{\frac{\pi}{2}} \varepsilon_\varphi(\lambda_i, T_i)\, 2\pi \sin\varphi \cos\varphi\, d\varphi}{\int_0^{\frac{\pi}{2}} 2\pi \sin\varphi \cos\varphi\, d\varphi}. \qquad (7)$$

From expression (7) the hemispherical spectral radiating power is obtained by averaging over the surface of the hemisphere.

Analogously, the normal total radiating power $\varepsilon_{\Sigma n}(T)$ may be defined as the ratio of the normal brightness of a polished nontransparent mirror made of the given material at a given temperature to the energy brightness of an absolutely black body at the same temperature.

The hemispherical total radiating power $\varepsilon_{\Sigma c}(T)$ equals the ratio of the luminances of a mirror made out of the given material and an absolutely black body at the same temperature. This $\varepsilon_{\Sigma c}(T)$ may also be determined by averaging, using an expression analogous to (7). In this expression instead of $\varepsilon_\varphi(\lambda_i, T_i)$ we need only put the relation for the total radiating power $\varepsilon_{\Sigma\varphi}(T)$ in terms of emission angle.

The $\varepsilon_c(\lambda_i, T)$ relationship for various materials was studied by several authors in the first quarter of the present century. The reader will find the relevant bibliography in collection [12].

Graphs of the spectral radiating power obtained by Worthing for tungsten at $\lambda = 0.665\ \mu$ are shown in Fig. 14.

TABLE 1

Emission angle, φ	Polarized light, %
0	0
30	10
45	22
60	46
75	72
80	81
85	90
90	100

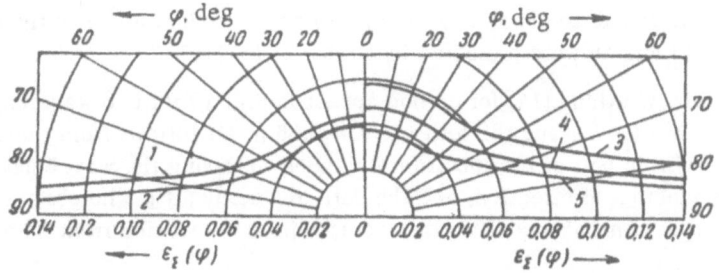

Fig. 17. Variation of total radiating power of some metals with emission angle.

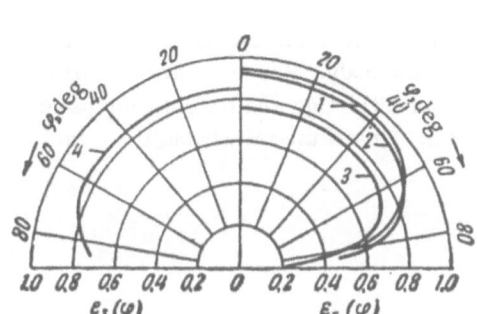

Fig. 18. Angular distribution of radiation for some substances.

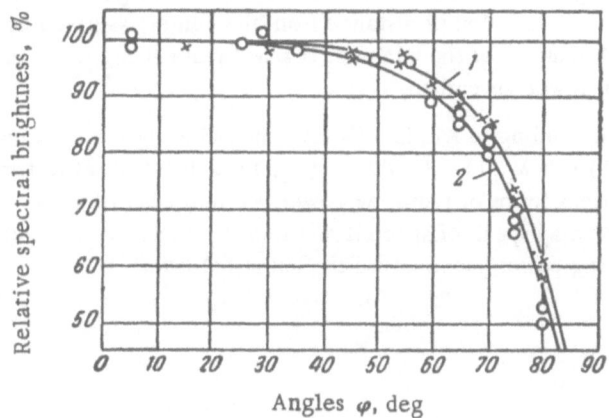

Fig. 19. Variation of spectral brightness for carbon: 1) For wavelength $\lambda = 0.47\ \mu$;   2) for wavelength $\lambda = 0.66\ \mu$.

Data obtained by the same author for the $\varepsilon_\varphi(\lambda_i, T)$ of tungsten at two wavelengths appear in Fig. 15.

The variation of the total and spectral radiating power (in the red part of the spectrum) for platinum is given in Fig. 16.

The proportion of polarized light in the total (natural) radiation of tungsten was determined quantitatively in experiments by Forsythe et al. [13]. These authors' data on the polarization of the radiation emitted by tungsten appear in Table 1.

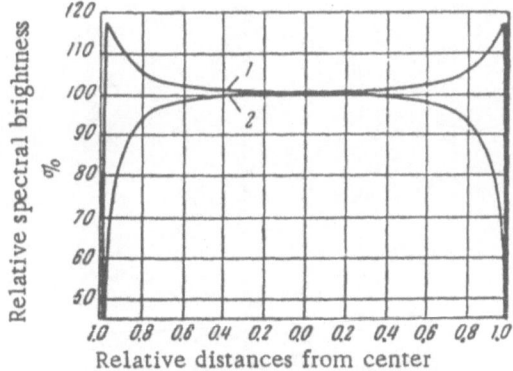

Fig. 20. Variation of spectral brightness ($\lambda = 0.66\ \mu$) along the diameter of an incandescent cylindrical filament. 1) For tungsten; 2) for carbon.

In working with tungsten filaments one must bear in mind the fact that the proportion of polarized light in a direction perpendicular to the axis is around 20% [9].

A very little oxidation of the surface of a metal mirror, however, creates conditions of diffuse radiation, and the distribution of radiating power in terms of the incident angle approaches a cosine law.

For illustration, we present the angular variation of $\varepsilon_\Sigma(T)$ for several metals in polar coordinates in Fig. 17. Here curve 1 gives the total radiating power of nickel for a polished surface and curve 2 that for a mat surface, while curves 3, 4, and 5 are for chromium, manganese, and aluminum, respectively.

From data of the same authors, Fig. 18 gives the total thermal-radiation distribution for the surfaces of several materials. In the cold state these materials may approximately be regarded as dielectrics.

Curve 1 of Fig. 18 corresponds to glass, 2 to loam, 3 to copper oxide, and 4 to the coarse surface of corundum. The data of Figs. 17 and 18 were presented by Gräber et al. [10].

The results obtained by Worthing [14] for carbon appear in Fig. 19 for two wavelengths. The upper curve corresponds to wavelength $\lambda = 0.47 \mu$ and the lower to $\lambda = 0.66 \mu$. It follows from these data that the deviation from Lambert's law in the visible part of the spectrum increases on raising the wavelength. The changes in the total radiating power as a function of direction, and the data of Czerny [15], who observed strong polarization in the infrared part of the spectrum, also confirm the increase in the deviation from Lambert's law with increasing wavelength.

Figure 20 illustrates the variation in the brightness of cylindrical incandescent filaments of tungsten and carbon as a function of distance from the center, as given by Ribault [9]. We see from these curves that the mean brightness of the filament is a few percent higher at the ends than at the middle for tungsten, and a few percent lower for carbon.

According to Kirchhoff's law, the radiating power equals the absorbing power. For a nontransparent body $\varepsilon(\lambda_i, T) + \rho(\lambda_i, T) = 1$, where (λ_i, T) is the spectral reflecting power of the body. Hence, in analogy with the various forms of radiating power, we may also imagine a like number of reflecting and transmitting powers, i.e., we may speak of spectral and total transmitting and reflecting powers. For diffuse reflection, the total reflecting power has received the name "albedo."

CHAPTER II

SOME THEORETICAL ASPECTS FROM THE PHYSICS OF THE RADIATION OF SUBSTANCES IN A CONDENSED PHASE*

For real bodies, radiating is always more or less selective. The variation of the radiating power with wavelength for real bodies emanates from fundamental aspects of the theory of dispersion and absorption, based on departments of electrodynamics concerned with the interaction of electromagnetic energy with matter. The frequencies of the natural oscillations of atoms and molecules of matter give a resonance character to the $\varepsilon = f(\lambda)$ relationship. For example, for metals, the region of anomalous dispersion (selective absorption of radiant energy) lies in the ultraviolet part of the spectrum. Hence the variation of radiating power with wavelength also here has a resonance character.

In accordance with the classical electron theory, the dispersion curves may be described [16] as

$$n^2 - k^2 = 1 + \sum \frac{4\pi m \left(\nu_0^2 - \nu^2\right) N e^2}{4n^2 m^2 \left(\nu_0^2 - \nu^2\right) + \gamma \nu^2},$$

where n, k are optical constants, e, m are the charge and mass of the electron, N is the number of electrons, ν_0 is the resonance frequency, and γ is the damping factor.

When the oscillations of all the electrons take place at frequency ν_0, electron theory leads to the following formula for the refractive index:

$$n^2 - 1 = \frac{4\pi e^2}{m_0} \sum_b \frac{N}{\nu_0^2 - \nu^2}.$$

On using quantum theory, when all the electrons are in the same state (a, for example), we have

$$n^2 - 1 = \frac{4\pi e^2}{m_0} \sum \frac{f_{ab} N}{\nu_{ab}^2 - \nu^2}.$$

Despite the fact that the general structures of the classical and quantum theories coincide, there are substantial differences between them. In the classical theory, the denominator contains the oscillation frequencies; in the quantum theory it contains "transition" frequencies (ν_{ab}). The values of these frequencies only coincide in the case of a harmonic oscillator. In the general case (anomalous dispersion), the conclusions of quantum theory are confirmed by experiment.

Another difference between the quantum and classical theories is that the number of scattering particles N is replaced by the oscillator strength $N f_{ab}$, and whereas N is a positive whole number $N f_{ab}$ may take negative fractional values. For the harmonic oscillator, as already indicated, $\nu_0 = \nu_{ab}$, and the quantum formula passes into the classical (allowing for the quantum-mechanical relation $\Sigma f_{ab} = 1$).

---

*The theoretical aspects presented in this chapter are set out in abbreviated form.

The theoretical calculations of function $\varepsilon(\lambda_i, T)$ for real bodies, in particular for metals, in the visible and near-infrared regions of the spectrum is beset by a number of serious difficulties. The solution of these requires a substantial development of the apparatus of quantum electrodynamics.

Use of the relations developed in papers of Krönig and Mott [17] and set out very fully in Sokolov's book [18], however, only leads to a qualitative estimate of the spectral distribution of radiating power for metals. In the far-infrared part of the spectrum, nevertheless, a satisfactory quantitative description of the function $\varepsilon(\lambda_i,T)$ for metals is given by Drude's well-known formula based on classical electron theory.

In view of the considerable interest of this, it is appropriate to give brief consideration to some fundamental aspects of physical optics.

Those desiring a more detailed treatment of such questions may refer to the corresponding Optics courses (Landsberg [1], Born [11]). As already mentioned, in accordance with Kirchhoff's law, for a nontransparent body

$$\varepsilon(\lambda_i, T) = 1 - \rho(\lambda_i, T).$$

The dependence of the reflection coefficient for normal incidence on the refractive index and absorption coefficient is given by the expression

$$\rho = \frac{n^2 + k^2 + 1 - 2n}{n^2 + k^2 + 1 + 2n}.$$

We remember that the quantities n and k are fundamental optical constants; methods for determining these independently may be found in the Optics courses cited.

Hence,

$$\varepsilon(\lambda_i, T) = 1 - \frac{n^2 + k^2 + 1 - 2n}{n^2 + k^2 + 1 + 2n} = \frac{4n}{n^2 + k^2 + 1 + 2n}.$$

On the basis of well-known electrodynamic relations [19], the optical constants n and k may be determined from the equations

$$2n^2 = \sqrt{E^2 + 4\frac{\sigma^2}{\nu^2}} + E,$$

$$2k^2 = \sqrt{E^2 + 4\frac{\sigma^2}{\nu^2}} - E,$$

where E is the dielectric constant, σ the electrical conductivity, and ν the frequency of the electromagnetic oscillations. For metals, in the long-wave part of the spectrum, the value of ν is small, and we may suppose $4\sigma^2/\nu^2 \gg E$, i.e., set $n = k = (\sigma/\nu)^{1/2}$.

Substituting the n and k values obtained into the earlier expression, we obtain for $\varepsilon(\lambda_i, T)$ the relation

$$\varepsilon(\lambda_i, T) = 2\left(\frac{\sigma}{\nu}\right)^{\frac{1}{2}} = 2\left(\frac{\nu}{\sigma}\right)^{-\frac{1}{2}}.$$

Replacing the electrical conductivity σ by the specific resistance r_0, measured in Ω/cm, and the frequency ν by the wavelength in μ, we obtain Drude's well-known formula in the following form:

$$\varepsilon(\lambda_i, T) = 36.05 \sqrt{\frac{r_0}{\lambda}}.$$

This expression shows that, on the basis of classical theory, the radiating power should rise with increasing specific resistance and fall with increasing wavelength.

Drude's formula gives satisfactory quantitative results for the majority of metals for wavelengths $\lambda > 10\,\mu$. For shorter wavelengths in the infrared part of the spectrum, Hagen and Rubens obtained the relation

$$\varepsilon(\lambda_i, T) = 0.365\left(\frac{r_0}{\lambda}\right)^{\frac{1}{2}} - 0.0667\frac{r_0}{\lambda} + 0.0091\left(\frac{r_0}{\lambda}\right)^{\frac{3}{2}}.$$

Here r_0 is the specific resistance in $\Omega \cdot mm^2/m$, λ is the wavelength in μ.

The three-term formula widens the possibility of the calculating the spectral radiating power of metals to approximately $\lambda > 5\,\mu$.

Using Planck's radiation law and Drude's formula for determining the $\varepsilon(\lambda_i, T)$ of a metal mirror, Ashkinas proposed a formula describing real radiation in the following form:

$$b_{met}(\lambda_i, T) = c_1 \cdot 0.365 r_0^{\frac{1}{2}} \lambda^{-5.5}\left(\exp\frac{c_2}{\lambda T} - 1\right)^{-1},$$

where $c_1 = 3.74041 \cdot 10^{-12}\,W \cdot cm^2$, $c_2 = 1.43868\,cm \cdot deg$.

Integrating this expression over the whole spectrum (from $\lambda = 0$ to $\lambda = \infty$), we obtain for the total radiation

$$b_\Sigma(T) = 12.27 c_1 \cdot 0.365 c_2^{-4.5} r^{\frac{1}{2}} T^{4.5}.$$

It is also known that for pure metals the specific electrical resistance depends to a first approximation linearly on temperature.

Introducing the notation r_{0y} for the specific resistance of 0°C and making the necessary calculations, we obtain the following expression for $b_\Sigma(T)$:

$$b_\Sigma(T) = 4.936 \cdot 10^{-20}\,c_1 r_{0y}^{\frac{1}{2}} T^5.$$

For pure metals the total radiation is proportional to the absolute temperature to the fifth power instead of the fourth as in the case of ideal radiation.

Experiments to Lummer and Kurlbaum and also Lummer and Pringsheim [20] showed that for platinum in the range 700 to 1800°K there is in fact a "fifth power" variation with temperature.

Figures 21-24 show results of measuring the radiating and reflecting powers of a metal mirror surface (aluminum and copper in Fig. 21, gold and nickel in Fig. 22, steel in Fig. 23). We see from the graphs for what parts of the infrared spectrum Drude's formula is valid. The continuous curves on these figures correspond

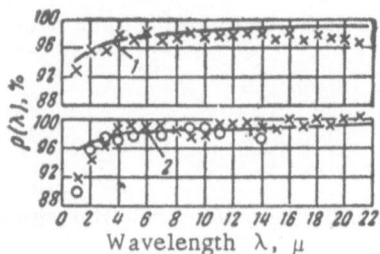

Fig. 21. Spectral reflecting power. 1) Aluminum mirror; 2) copper mirror (continuous lines correspond to theoretical calculations of Snyder [21] from Drude's formula,
circles indicate experimental data of Hagen
and Rubens, crosses indicate experimental
data of Gier, Dunkel, et al.).

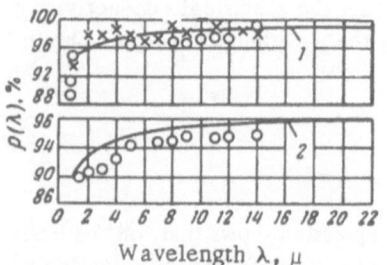

Fig. 22. Spectral reflecting power. 1)
Gold; 2) nickel (continuous lines correspond to theoretical calculations of
Snyder [21], circles indicate experimental data of Hagen and Rubens, crosses indicate experimental data of Gier, Dunkel,
et al.).

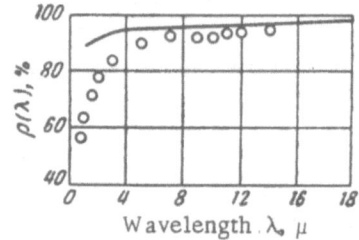

Fig. 23. Reflection of steel mirror (continuous curve theoretical, circles give
Hagen and Rubens' experiments).

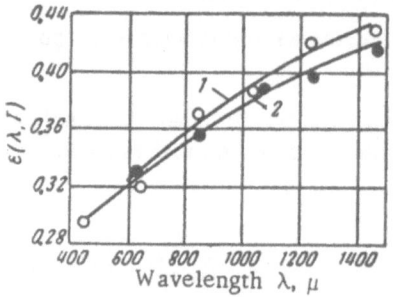

Fig. 24. Spectral radiating power of
platinum as a function of temperature.
1) For wavelength $\lambda = 4\mu$; 2) for wavelength $\lambda = 6\mu$.

to the theoretical values calculated by Drude's formula, while the circles indicate the experimental data of
Hagen and Rubens. Figures 21-23 characterize the radiating power as a function of wavelength at room temperature. The crosses on Figs. 21 and 22 indicate the results of experiments by Gier, Dunkel, et al.[40]. Figure 24
shows the variation of the radiating power of platinum with temperature for wavelengths of 4 and 6 μ. The light
and dark circles on this figure correspond to experimental data obtained by Hagen and Rubens for wavelengths
of 6 and 4 μ, and the continuous curves are constructed from Drude's formula.

For every pure metal there is a wavelength [the so-called "X point" on the $\varepsilon(\lambda)$] for which

$$\left| \frac{\partial \varepsilon(\lambda_i, T)}{\partial T} \right|_{\lambda = \lambda_\chi} = 0,$$

i.e., the radiating power is independent of temperature.

For the majority of metals which have been studied, the wavelength λ_χ for which $\varepsilon(\lambda_i, T)$ has a zero
temperature coefficient lies in the near-infrared part of the spectrum. Thus for tungsten (see below) $\lambda_\chi = 1.3\,\mu$,
for rhenium $\lambda_\chi = 1.4\,\mu$. A special study of the temperature coefficient of radiating power for metals was made
by Price [22] and Ward [23].

The direction of the temperature dependence of the spectral radiating power of a metal in the infrared part of the spectrum is determined by the change in the electrical resistance with temperature. As a temperature rises, the radiating power of metals in the infrared spectrum does likewise.

Application of classical theory to the visible and ultraviolet parts of the spectrum, i.e., in the region of anomalous dispersion, as indicated before, is inappropriate for quantitative determinations of radiating power. It has been established experimentally that in these spectral regions the temperature coefficient of radiating power for a number of metals is negative.

In general, Drude's formula, based on classical theory, is in principle applicable to metals, giving satisfactory agreement with experiment, for certain ranges of wavelength only: for tungsten, $\lambda > 5 \mu$, for molybdenum, $\lambda > 10 \mu$, and so forth. Thus Drude's formula and Ashkinas' transformation can only be of interest for the medium- and long-wave parts of the spectrum.

One of the reasons for this is the assumption made in deriving Drude's formula that Ohm's law is valid for the current induced in the metal by the electromagnetic field of the incident light wave. But Ohm's law will only be valid if the period of the frequency of the incident radiation is not too small in comparison with the free path of an electron in the metal.

For the shorter-wave region, classical concepts, in particular the classical electron theory, on which Drude's formula is based, are inapplicable. This is clearly illustrated by the isotherms of $\varepsilon(\lambda_i, T_i)$ for tungsten in Fig. 25, taken from Ribault [9], as obtained by experiment (continuous curves) and as calculated by Drude's formula (broken curves). As we see, even in the range $\sim 1 \mu$ the discrepancy between experiment and classical theory is hundreds of percent.

The weak dependence of the spectral radiating power on temperature leads to a slight deviation between the characters of the spectral distribution of $\varepsilon(\lambda, T)$ for a liquid and solid metal mirror.

This is evidently connected with the fact that, in a metal liquid, "short-range order," which mainly determines the variation of the quantity with wavelength, is preserved. The sharp rise in brightness ("flash") which may be observed on solidification of a metal bath with a clean (unoxidized) surface is most probably due to the fact that the surface becomes rough on solidification owing to the formation of grains (crystals).

This phenomenon was studied by Svet and Talenskii [24] with oscillograph recording of the intensity of radiation from the unoxidized surface of a crystallizing Armco-iron bath (argon atmosphere).

In all parts of the visible spectrum, at the moment of solidification the oscillogram recorded a "splash" of brightness. Quantitatively, the increase in brightness and hence also the value of $\varepsilon(\lambda, T)$, depending on the conditions of crystallization, may reach 100%. Here it should be noted that the effect of the crystal structure of the previously polished surface of a metal sample on the value of radiating power may be considerably distorted by an oxide film, which may not be very noticeable. As an example we may indicate the results of new investigations by Butler and Edward [25] in measuring the hemispherical total radiating power $\varepsilon_\Sigma(T)$ for polished copper samples. The measurements in this work were made calorimetrically on heating the sample under examination by radiant heat through a special window. For cooling, water and liquid nitrogen were used.

Figure 26 shows a microphotograph of the surface of a polished sample after five-times heating at 900°C. After chemical etching, another microphotograph was taken (Fig. 27), and on this the grain boundaries are very clearly visible.

Data on the hemispherical total radiating power $\varepsilon_\Sigma(T)$ of this sample, on heating to not more than 550°C, are shown in Fig. 28 as curve 1 (black circles for measurements made with liquid nitrogen). In this figure, curve 2 corresponds to measurements of the hemispherical total radiating power $\varepsilon_\Sigma(T)$ of the surface of the same sample, but with structure "developed" as a result of many heatings to 990°C in vacuum. The microphotograph of this surface appears in Fig. 29.

Despite the apparent similarity in the structure of the two samples, the radiating power of the second is considerably smaller. This difference is due to the influence of surface oxidation in the first sample during chemical etching.

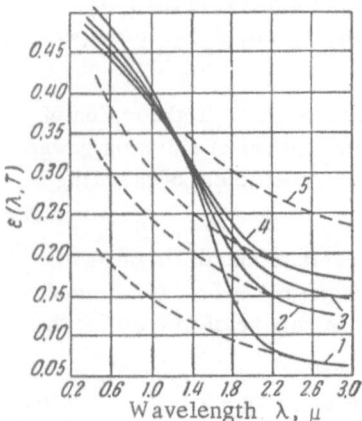

Fig. 25. Spectral radiating power of tungsten as a function of λ for various temperatures expressed in °K (continuous curves experimental, broken curves calculated by Drude's formula). 1) for 300°C; 2) for 1300°C; 3) for 1700°C; 4) for 2100°C; 5) for 3500°C.

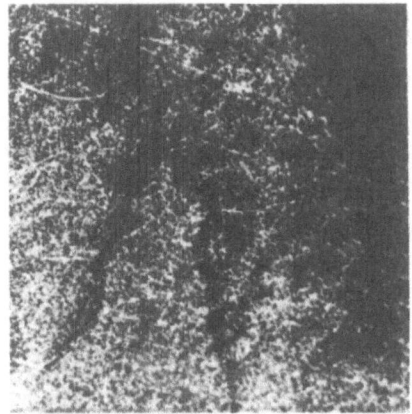

Fig. 26. Microphotograph of copper after five-times heating to 900°C.

Fig. 27. Microphotograph of surface of Fig. 26 after chemical etching.

Because the first sample was not heated above 550°C, the oxide film was not reduced.

Radiating Power of Dielectrics

The quantitative theory of the radiation of dielectrics, in particular the oxides of metals, in the optical region of the spectrum has even at the present time not been developed sufficiently for practical calculations.

The characteristic oscillation frequencies of the ions lie in the infrared region. For radiation in the visible region, the refractive index, determined by Maxwell's theory [19] as the square root of the dielectric constant, varies with the wavelength; this is connected with the gradual approach to the characteristic oscillation frequencies of the electrons, the values of which lie in the ultraviolet region of the spectrum.

On the basis of the well-known Frésnel formulas and the relations of classical electrodynamics given above, the reflection coefficient is connected with the refractive index by the expression

$$\rho(\lambda) = \left(\frac{n-1}{n+1}\right)^2.$$

For subsequent calculation of $\varepsilon(\lambda)$ in transparent dielectrics, the dispersion formula which was written down for metals simplifies, since here the term characterizing damping will be absent.

In the form of the following expression, this formula is known in physics as Sellmeier's dispersion formula

$$n^2 - 1 = \sum_k \frac{A_k}{\nu_k^2 - \nu^2}. \tag{8}$$

For a nontransparent dielectric, on the basis of Kirchhoff's law

$$\varepsilon(\lambda) = \frac{4n}{(n+1)^2}.$$

The change in the radiating power of the dielectric with temperature is connected with the change in its volume resulting from thermal expansion.

Let us use the relation employed in modern solid-state physics between the refractive index and temperature in the form

$$n = \left[1 + \frac{cN_0}{1 + a(T - T_0)}\right]^{\frac{1}{2}}, \tag{9}$$

where a is the expansion coefficient, N_0 the number of oscillators in unit volume at temperature T°, and c = const.

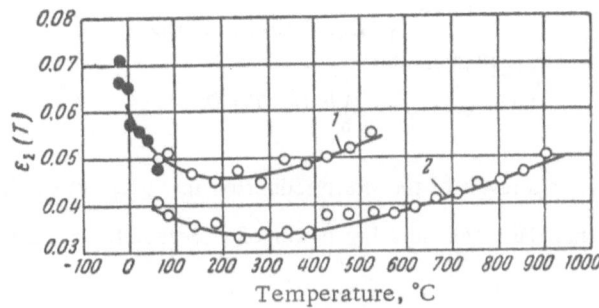

Fig. 28. Variation of the hemispherical total radiating
power of copper with temperature.

After due transformations, we obtain the expression given in a paper by Bevans et al. [26] for the temperature dependence of the radiating power of a dielectric:

Fig. 29. Microphotograph of the surface
of copper after repeated heating to 990°C
in vacuum.

$$\frac{\partial \varepsilon (\lambda_l, T)}{\partial T} = \frac{2cN_0 \left\{ \left[1 + \frac{cN_0}{1 + a(T - T_0)} \right]^{\frac{1}{2}} - 1 \right\}}{[1 + a(T - T_0)]^2 \left\{ \left[1 + \frac{cN_0}{1 + a(T - T_0)} \right]^{\frac{1}{2}} + 1 \right\}^{3,5}} \cdot \tag{10}$$

This formula only qualitatively represents the situation.
For example, it shows that on raising the temperature the spectral radiating power also rises.

However, the well-known fact of the fall in the total radiation power of dielectric on raising the temperature is not made
clear by this formula.

Together with the authors of [26] and some other investigators, we may suppose that the reason for this kind of temperature dependence of the total radiating power is the displacement
of the radiation maximum on increasing the temperature, in
accordance with the Planck-Wien law, to the short-wave side,
where $\varepsilon(\lambda_i, T)$ diminishes.

A graph from the paper of Bevans et al. [26] illustrating
this situation, finding some support in [29] (considered below),
appears in Fig. 30. Here $T_2 > T_1$ and the scales along the ordi-
nate axes are arbitrary.

From this figure we may gather that, despite the growth in
the monochromatic radiating power, on raising the temperature
$\varepsilon(\lambda_i, T_2) > \varepsilon(\lambda_i, T_1)$, and the total radiating powers, defined as

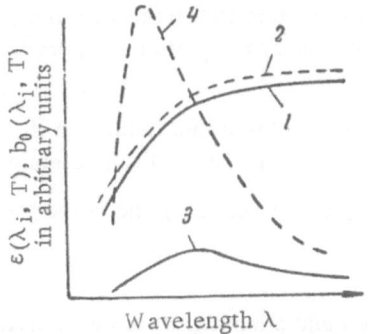

Fig. 30. Hypothetical curves of $\varepsilon(\lambda, T)$
explaining expression (10). 1) $\varepsilon(\lambda_i, T_1)$;
2) $\varepsilon(\lambda_i, T_2)$; 3) $b_0(\lambda_i, T_1)$; 4) $b_0(\lambda_i, T_2)$.

$$\varepsilon_\Sigma (T_1) = \frac{\int\limits_0^\infty \varepsilon (\lambda_l, T_1) \, b_0 (\lambda_l, T_1) d\lambda}{\int\limits_0^\infty b_0 (\lambda_l, T_1) \, d\lambda},$$

27

$$\varepsilon_\Sigma\,(T_2) \coloneqq \frac{\int\limits_0^\infty \varepsilon\,(\lambda_l,\,T_2)\,b_0\,(\lambda_l,\,T_2)\,d\lambda}{\int\limits_0^\infty b_0\,(\lambda_l,\,T_2)\,d\lambda}\,,$$

will correspondingly fall faster, as a result of the sharp reduction in $\varepsilon(\lambda_i,\,T)$ on shortening the wavelength.

Thus we may conclude that the total radiating power of a dielectric will fall on raising the temperature, i.e., $\varepsilon_\Sigma(T_2) < \varepsilon_\Sigma(T_1)$.

As already indicated, the radiating power is a constant of the surface, and not the substance.

For polycrystalline oxides and various kinds of ceramic materials, the values of $\varepsilon(\lambda_1,\,T)$ and $\varepsilon_\Sigma(T)$ will depend not only on the chemical composition but also on the density and granulometric composition (grain size). Clearly, these factors, for a given type of treatment, will finally determine the microgeometry of the emitting surface of the polycrystalline material, and the functional relationships obtained by various authors for the radiating powers $\varepsilon(\lambda,\,T)$ and $\varepsilon_\Sigma(T)$ must be approached with great caution. This is associated both with the very concept of the "purity of treatment" of the surface and also with the possible methodical errors arising in measurements. In some papers, including a broad experimental study [27], the results obtained are for these reasons subject to doubt in the qualitative as well as the quantitative respect.*

A very frequent source of methodical error in such measurements is the existence of temperature gradients when measuring the radiating powers from the radiation (see below), owing to the low thermal conductivity of the oxides, and the different degrees of roughness of the surface.

As a result of extremely circumstantial investigations of radiating power in refractory materials by Michaud [28], supported by later careful experiments and calculations of Allegre [29], we may conclude as follows. The total radiating powers $\varepsilon_\Sigma(T)$ of refractory materials are usually equal to or greater than their radiating powers $\varepsilon(\lambda,\,T)$ for the red part of the spectrum. This indicates the existence of strong emission (absorption) of these materials in the infrared part of the spectrum, which is fully confirmed by spectral examination. The $\varepsilon_\Sigma(T)$ coefficients of refractory materials always fall on increasing the temperature, owing to the displacement of the radiation maximum, on the basis of Wien's law, in the shorter-wave region, where the $\varepsilon(\lambda,\,T)$ of refractory materials falls. For silicon—aluminum refractories of similar grain size, the spectral radiating powers depend very little on the silicon and aluminum content. Individual differences in the granulometry of refractory material may be associated with a difference in the character of the $\varepsilon(\lambda,\,T)$ function. There is no systematic influence of the ratio between SiO_2 and Al_2O_3 on the value of the spectral radiating power.

In general, the value of the radiating power of refractory materials is determined by the absorption and scattering of radiant energy in these. The absorption coefficient depends on the optical constant n, the extinction index of the crystals, the material of the refractory, and its porosity. The scattering coefficient in turn is determined by the reflection of the crystals, the diameter of their particles, and the porosity. Thus the thermal radiation of refractory materials may be characterized by means of parameters associated with the optical constants of its component elements and with geometric factors determined by the grain size, porosity, and the treatment of the surface.

These concepts are very convincingly confirmed by studies of Allegre [29]. This author compared the reflection of a single crystal and a polycrystalline refractory at ordinary and high (up to 1000°C) temperatures.

Allegre's [29] curves for the reflection coefficient of a quartz single crystal (broken line) and polycrystalline silicon oxide refractory (continuous curve) are shown in Fig. 31.

---

*In the paper in question [27], contrary to the fundamental theoretical position and a number of circumstantial modern investigations, a rise in the total radiating power of oxides on increasing the temperature and so forth was observed.

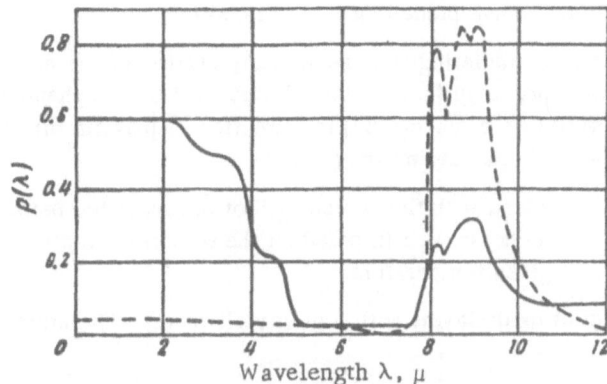

Fig. 31. Reflecting power of a quartz single crystal (broken line) and polycrystalline SiO_2 (continuous curve) according to Allegre [29].

The composition of the refractory was as follows: 95% SiO_2, 3% CaO, and 2% kaolin. Briefly, the interpretation of these curves is that, in the spectral range up to 3 μ, the reflection coefficient $\rho(\lambda)$ of the single crystal may be determined from the refractive index n by way of expression $\rho(\lambda) = (n - 1/n + 1)^2$. Here the reflection coefficient is small and falls with increasing wavelength.

Beginning from around $\lambda = 3 \mu$, the absorption rises and reaches 100% for $\lambda = 6 \mu$. This takes place as a result of an increase in the extinction coefficient, which reaches a maximum at $\lambda = 8.84 \mu$.

The part of the spectrum from 7 to 12 μ (the so-called "metallic-reflection" band) characterizes the anomalous-dispersion region of quartz.

In this region, the $\rho(\lambda)$ relationships for the single crystal and polycrystalline refractory coincide.

The lower reflection coefficients for the single-crystal quartz may be explained by diffuse scattering. It must be noted that the small reflection coefficient of single-crystal quartz up to 3 μ characterizes high transmission and not absorption.

For the polycrystalline refractory, however, the apparent reflection coefficient in this region of the spectrum increases as a result of multiple scattering and reflection.

In the region of greatest absorption, the curves of $\rho(\lambda)$ for the single crystal and polycrystalline refractory coincide.

Also extremely important are the processes of transmitting electromagnetic energy and thermal radiation by layers of various oxides. At ordinary temperatures, questions connected with the transmission coefficients of various materials are mainly of interest from the point of view of use as optical components (windows, lenses, etc.).

Let us briefly consider the radiating power of the surfaces of several oxides allowing for their transmission coefficients.

Oxides of both polycrystalline and vitreous varieties are of practical interest. Naturally, for polycrystalline oxides, questions associated with their transmitting power are only of interest for fairly thin layers. At the same time, for glasses, only by allowing for transmission processes, as indicated earlier (McMahon's theory), can we make a correct estimate of the radiating power.

The transmission of polycrystalline quartz was considered by Allegre in the paper already mentioned [29].

Naturally, this kind of investigation requires extremely thin plates. In view of the experimental difficulties arising in producing these, owing to the grain size of the refractories, the work was effected with powders.

In all cases, as well as establishing the absorption bands characterized by anomalous dispersion, an increase of transmission in the spectral range preceding 8 μ was observed.

This phenomenon, apparently contradicting the above interpretation of the $\rho(\lambda, T)$ curve illustrated in Fig. 31 for this polycrystalline refractory, is in fact explained very well by the phenomenon of scattering in the layer of powder. The powder particles were placed [29] in a solution of paraffin oil. This mass "in layer" constitutes a peculiar "light filter" of the Christiansen type [1, 11].

For grain sizes fairly small compared with the wavelength of the radiation being studied and of the same order as the oxide layer, scattering effects are superimposed on the selective-transmission characteristic due to the properties of the oxide itself at a given temperature.

The spectral radiating power of oxide layers with a grainy character will naturally also depend on the granulometric composition.

A thin film of such oxide, formed, for example, on the polished surface of a metal, may substantially raise the absolute value of the spectral and total radiating power, not only as a result of its "own" optical constants, but also owing to the development of "roughness" on the surface of the metal.

Radiating Power of Glasses

As a result of the fall in absorption from scattering, oxides in the vitreous state are distinguished from those in the polycrystalline state by higher transmission coefficients.

Expressions for the radiating and transmitting powers of partially transparent materials (McMahon theory) were considered earlier. Fundamental to this theory is the calculation of $\varepsilon^*(\lambda, T)$ with due allowance for multiple reflections inside the plate.

In the course of the McMahon discussion, the very important concept of the spectral volume radiating power of the material was introduced.

In McMahon's conclusions, however, there are two substantial assumptions: The first concerns fluxes of radiation normal to the glass surface, i.e., in essence McMahon's formula does not extend to the hemispherical radiating and transmitting powers; the second is that McMahon's theory does not consider questions connected with the refractive index.

Certain questions connected with heat transfer were solved by Kellett [30].

Czerny [31] showed that in general the problem of heat transfer in partly transparent bodies must be considered as a three-dimensional problem even for directional radiation.

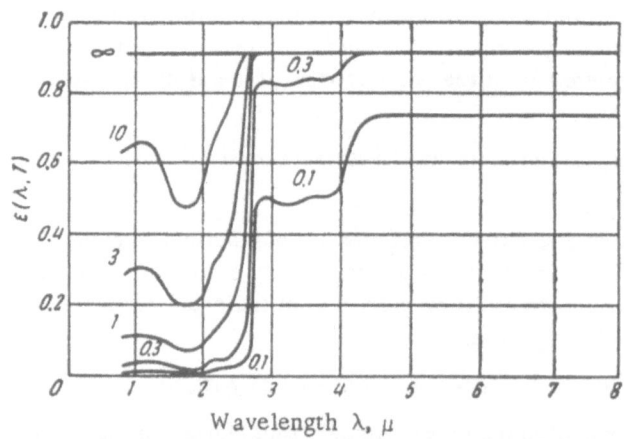

Fig. 32. Hemispherical $\varepsilon(\lambda, T)$ of glass at 1000°C for various layer thicknesses (from 0.1 to 10 cm).

TABLE 2

Thickness, cm	Total radiating power $\varepsilon_\Sigma(T)$ at 1000°C
0.1	0.4
0.3	0.57
1.0	0.63
3.0	0.82
10.0	0.91

No specific solution of the problem of radiating power, however, was reached in this paper. The question of the thermal radiation of transparent materials and the temperature distribution in glass plates was dealt with in some detail in papers by Gardon [32, 33].

Data on the hemispherical spectral radiating power of glass as a function of layer thickness [32] at 1000°C are given in Fig. 32.

The variation of total radiating power with layer thickness is shown in Table 2.

The total hemispherical radiating power of glass as a function of layer thickness and temperature appears in Fig. 33.

All these data were obtained in the paper cited with due allowance for the reflection coefficient and refractive index for an isothermal layer of glass.

As a result of poor thermal conductivity, however, thermal radiation by a layer of semitransparent material, especially glass, may be complicated by the existence of a temperature gradient over its thickness. The question of this phenomenon is extremely important, especially for the pyrometry of material partly transmitting thermal radiation, and in particular for glass.

In fact, the use of any method of radiation pyrometry for measuring the temperature of a partly transparent material will be complicated by the fact that energy is emitted not only from its surface but from the inside, which in general will have a temperature differing from that of the outer surface.

In practice, the main part in measuring the surface temperature of a plate by the radiation method is played by the normal radiating power, since, as a rule, it is the normal radiation which is received by the pyrometer.

Beattie [34] studied the normal radiating power of a glass plate having either a constant temperature or a temperature varying through its thickness. For isothermal, linear, and parabolic laws of temperature distribution through the thickness of a layer, the expression for the spectral power radiated normally by the glass plate has the following form:

$$E(\lambda, T) = [1 - \rho(\lambda, T)] \frac{\alpha(\lambda) c_1}{\pi \lambda^5} \int_0^d \exp\left(-\frac{c_2}{\lambda T}\right) \exp[-\alpha(\lambda) x] dx +$$

$$+ \rho(\lambda, T) \exp[-\alpha(\lambda) d][1 - \rho(\lambda, T)] \frac{\alpha(\lambda) c_1}{\pi \lambda^5} \int_0^d \exp\left(-\frac{c_2}{\lambda T}\right) \exp[-\alpha(\lambda) x] dx. \quad (11)$$

The condition $T = T_0$ for $0 \le x \le d$ corresponds to constant temperature over the glass layer; $T = T_0 = ax$ for $0 \le x \le d/2$ and $T = T_{max} - a(x - d/2)$ for $d/2 \le x \le d$ to the linear temperature distribution; $T = T + ax^2 + bx$ for $0 \le x \le d$ to the parabolic temperature distribution.

Here we have used the following notation: $T_0 =$ surface temperature of the glass, °K; $T_{max} =$ maximum temperature along the central line of the glass, °K; $\alpha(\lambda) =$ spectral absorption coefficient of the glass for wavelength λ, μ; $\rho(\lambda) =$ reflection coefficient of the surface for wavelength λ, μ; $d =$ thickness of glass, cm.

The first term equation (11) is the direct radiation coming from the surface of the glass and passing into the atmosphere.

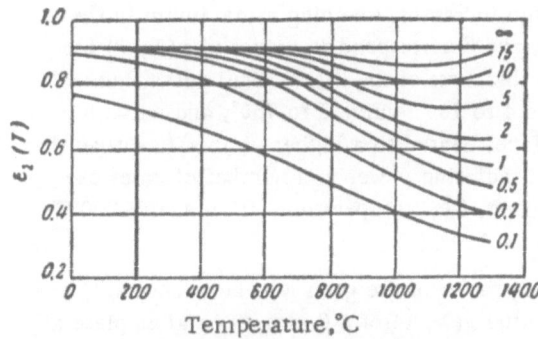

Fig. 33. Total hemispherical radiating power of glass as a function of temperature for various layer thicknesses (from 0.1 to 15 cm).

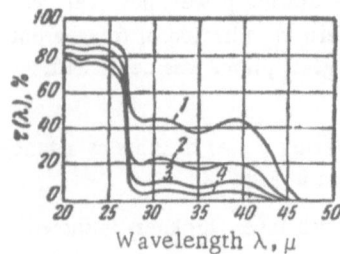

Fig. 34. Spectral transmission of glass layers of various thickness: 1) 1.6 mm; 2) 3.2 mm; 3) 6.4 mm; 4) 20 mm.

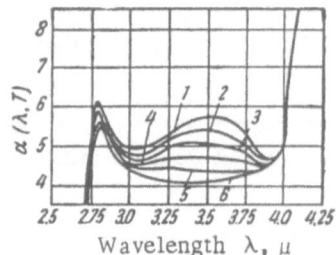

Fig. 35. Spectral absorption of a glass layer at various temperatures: 1) 200°C; 2) 300°C; 3) 400°C; 4) 500°C; 5) 600°C; 6) 700°C.

The second term is the radiation reflected from the back surface of the glass, returning to the front surface and then issuing into the atmosphere.

Ordinary glass (main components: soda, calcium oxide, silicon) becomes nontransparent for wavelengths $\lambda > 5\ \mu$ (Fig. 35). For example, for $\lambda > 5\ \mu$, $\alpha(\lambda)$ is infinitely large, and hence $\exp[-\alpha(\lambda)d] = 0$. In the range 2.75 to 4.5 μ, $\rho(\lambda)$ has a value approximately equal to 0.04, and $\exp[-\alpha(\lambda)d]$ is 0.3 and 0.09 for glass of thickness 3 and 6 mm, respectively. Thus the maximum value of the product $\rho(\lambda)\exp[-\alpha(\lambda)d]$ will be ~ 0.012. Hence all internal reflections except that of the primary radiation from the back surface of the glass, which has the comparatively high value $\rho(\lambda)\exp[-\alpha(\lambda)d]$, may be neglected.

The solution of the equation is complicated by the fact that inside the wavelength range 2.75 to 4.5 μ, $\alpha(\lambda)$ depends not only on the wavelength of the radiation but also on the temperature of the glass (Fig. 35); the figure shows the absorption of the glass in cm^{-1} as a function of temperature.

To simplify calculations, the values of $\alpha(\lambda, T)$ obtained for glass at constant temperatures 300 and 600°C over the whole thickness may be used when the presumptive surface temperatures lie in the ranges 200 to 400 and 500 to 1000°C, respectively.

The power of the spectral radiation obtained from equation (11) for plane glass 3 and 6 mm thick with surface temperature 600°C appears in Figs. 36 and 37, respectively.

In Fig. 36, curve 1 corresponds to an isothermal temperature distribution over the thickness of the plate, curve 2 to a linear distribution with maximum temperature 650°C. Curves 3, 4, and 5 correspond to a parabolic temperature distribution over the thickness of the glass plate. The maximum temperatures are 625° for curve 3, 650° for curve 4, and 700°C for curve 5.

The notation in Fig. 37 is analogous. The extra curves 6 and 7 here correspond to a parabolic distribution with maximum temperatures of 750 and 800°C, respectively.

Curves of the spectral radiating powers of glass are shown in Figs. 38 and 39. These curves were obtained from the graphs of the spectral radiation power for a surface temperature of 600°C, for thicknesses 6 and 3 mm, respectively, and various temperature gradients, relative to the spectral radiating power of a black body at the same temperature as the glass surface.

In both figures, curves 1 correspond to an isothermal temperature distribution. Clearly, with increasing wavelength the value of $\varepsilon(\lambda, T)$ for glass in this part of the spectrum increases, reaching a maximum in the region of 5 μ, where the glass is nontransparent. Curves 2, 3, 4, 5, and 6 correspond to values of $\varepsilon(\lambda, T)$ for various gradients and a parabolic temperature distribution over the thickness of the glass. Thus curve 2 corresponds to a maximum temperature of 625°, curve 3 to 650°, curve 4 to 700°, curve 5 to 750°, and curve 6 to 800°C. It is clear that, as a result of radiation from the "back" surface, heated to a higher temperature than the front, which in all cases was heated to 600°C, the resulting spectral radiating power in a number of cases exceeds 100%. In particular, for $T_{max} = 800°C$ and film thickness 6 mm, i.e., temperature difference of 200°C between the front and back surfaces of the plate $\varepsilon(\lambda, T)$ will reach ~ 190% (!).

In the region of "total nontransparency," the spectral radiating power of the glass will be accordingly determined by its reflecting power. Curve 1 of Fig. 40 corresponds to the $\varepsilon(\lambda, T)$ of a 6 mm-thick glass plate at 300°C, and curve 2 to 600°C; curves 3 and 4 correspond to a 3-mm plate at the same temperatures. At the foot of the same figure is the curve for $\rho(\lambda, T)$, the spectral reflecting power of the glass plate.

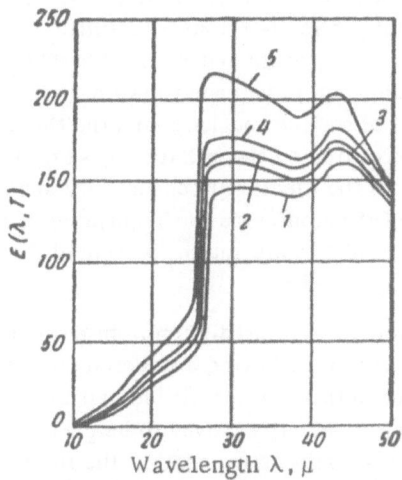

Fig. 36. Spectral energy emitted normal to a layer of glass 3 mm thick for various temperature-gradient distribution laws.

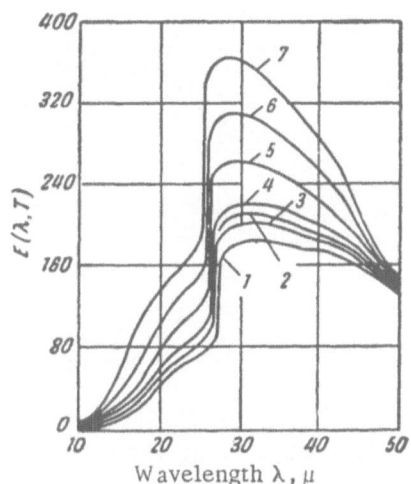

Fig. 37. Spectral energy emitted normal to a layer of glass 6 mm thick for various temperature distributions over the thickness of the layer.

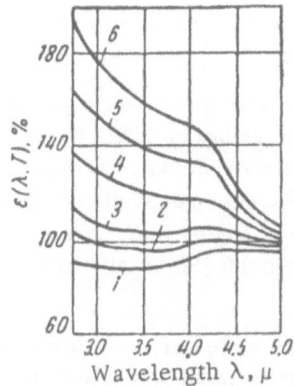

Fig. 38. Spectral radiating power of a glass plate 6 mm thick relative to a black body at 600°C and various temperature distributions over the thickness of the layer.

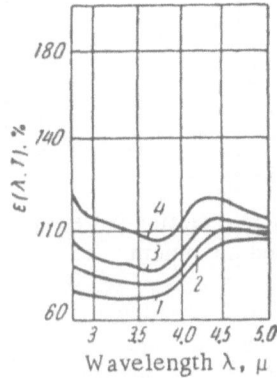

Fig. 39. Spectral radiating power of a glass plate 3 mm thick relative to a black body at 600°C and various temperature distributions over the thickness of the layer.

Radiation and Reflection of Thin Layers (Films)

Naturally, the emission of radiation from thin transparent films may be accompanied by interference. Theoretically, interference may seriously distort the picture of the spectral relationships $\varepsilon(\lambda, T)$ or $\rho(\lambda, T)$ of a metal mirror if the surface is covered by a very thin oxide layer. It is enough to recollect the temper color on a polished and oxidized steel surface, the golden color of anodized aluminum, and so forth.

It is very risky, however, to draw conclusions regarding the selective radiation of this or that oxide from its properties at room temperature. This is because, as temperature rises, not only does the electrical conductivity (determining the coefficients of transmission, radiating power, and reflection) change, but the actual structure of the film alters.

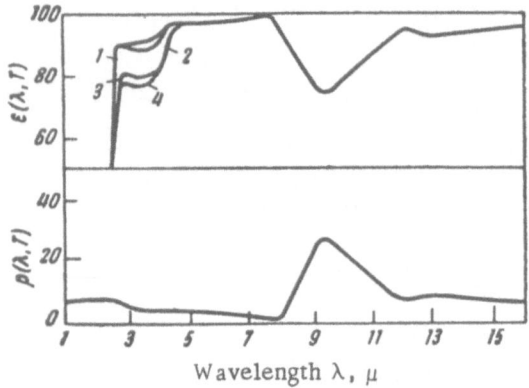

Fig. 40. Spectral radiating power.

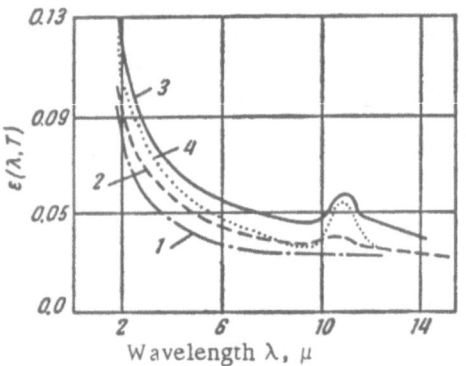

Fig. 41. Spectral radiating power of polished aluminum

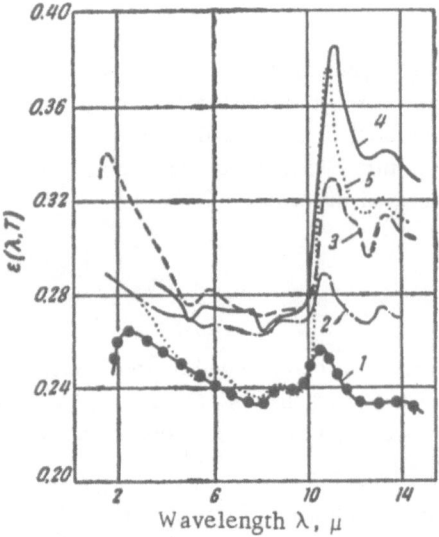

Fig. 42. Spectral radiating power of unpolished aluminum.

In certain papers of a speculative nature, failure to allow for this has led to erroneous results. An example is the article of B. V. Stark and Yu. M. Shashkov [35]. In this paper an attempt is made to estimate by calculation the effect of surface roughness and the thickness of the oxide film of iron on the radiating power. The absence of data on the electrical conductivity and dielectric constant of the iron oxide for frequencies of the optical range at high temperatures prevented the problem from being solved correctly.

Thus, instead of the radiation from the oxide film on the iron surface having a selective nature, as proposed by the authors, in the visible part of the spectrum the radiation actually proved to be gray, and so on. Here we must specially notice that, as the thickness of the film changes, so does its roughness.

Moreover, the film on the surface of the metal in the cold state, which produces interference of the reflected light, not only changes its transparency on heating, as a result of changes in the electrical conductivity, dielectric constant, and degree of roughness, but also (and this is very fundamental on heating in air) alters its thickness with temperature.

It is thus extremely important in practice to examine changes in the values of $\varepsilon(\lambda, T)$, $\rho(\lambda, T)$ for an oxidized surface in the dynamics of various technological processes (see below). As an example of an oxide film remaining stable for changing temperature we may quote the data of Reynolds [36] on the $\varepsilon(\lambda, T)$ of aluminum.

Here Fig. 41 shows the $\varepsilon(\lambda, T)$ of a polished aluminum surface in the infrared part of the spectrum. The characteristic flash around $\lambda = 11\,\mu$ owes its origin to an interference phenomenon due to the presence of the oxide film. The increase in the spectral radiating power on raising the temperature takes place as a result of oxide-film growth (naturally oxidation increases as the temperature rises). In this figure, curve 1 corresponds to heating for 19 h at 326 ± 4°C, curve 2 to 20 h at 424 ± 4°C, and curve 3 to 15 h at 535°C. From the latter, the temperature was again reduced to 326°C (curve 4).

As we see, during these heating processes the value $\varepsilon(\lambda, T)$ rose as a result of oxidation. Data on $\varepsilon(\lambda, T)$ obtained in the same investigation for a rough aluminum surface are shown in Fig. 42. The roughness substantially raises the value of $\varepsilon(\lambda, T)$. Curve 1 for 190°C, 2 for 330°C, 3 for 440°C, and 4 for 535°C as before characterize the growth of $\varepsilon(\lambda, T)$ on raising the temperature.

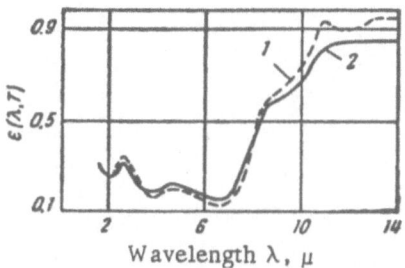

Fig. 43. Spectral radiating power of
an anodized aluminum surface.

Repeated cooling with holding up to 190°C (curve 5) characterizes a growth in $\varepsilon(\lambda, T)$ in the region $\lambda > 10 \mu$ as a result of oxidation.

Figure 43 shows two curves: 1 for 188°C and 2 for 300°C, for anodized aluminum. As we see, for a stable film obtained by anodizing, no change in $\varepsilon(\lambda, T)$ with temperature is observed. Figure 43 clearly illustrates the effect on the spectral radiating power of interference phenomena due to the thin surface film resulting from anodization.

In the example given, the oxide film (Al_2O_3) is extremely stable, and does not collapse in the temperature range used (up to 500°C).* In evaluating the results of measuring the $\varepsilon(\lambda, T)$ of a metal surface in the presence of oxide films, we must always bear in mind the effect of experimental conditions, not only from the quantitative but also the qualitative point of view.

In order to confirm what we have said, the reader may compare the above $\varepsilon(\lambda, T)$ curves for oxidized aluminum with the results (given later) for the same metal taken from the well-known paper by Hase [37]. It may also be noted that in industrial conditions the change in the $\varepsilon(\lambda, T)$ of the oxidized surface of aluminum alloys for a narrow range of the near-infrared part of the spectrum ($1.6 \mu < \lambda < 2.0 \mu$) gives full-reproducible results without any interference effects. The absence of these is evidently due to the considerable thickness of the oxide film.

Here we must also bear in mind the fact that, from the point of view of diffraction phenomena, the scale of the film "thickness" is the wavelength of the radiation being studied. Thus one and the same film is "thicker" with respect to the visible region than the infrared.

In practice, interference phenomena from liquid films on the surface of a metal bath play a still smaller part in the visible region of the spectrum.

The reason for this phenomenon lies not only in the high electrical conductivity of molten slags but also in their anisotropy.

Changes in the radiating power of metal surfaces covered by oxides will be considered below for several specific cases, in the dynamics of film development.

---

*Al_2O_3 is an excellent refractory material up to temperatures of about 1800°C.

CHAPTER III

METHODS OF DETERMINING SPECTRAL AND TOTAL
RADIATING-POWER COEFFICIENTS

In the visible and near-infrared regions of the spectrum there is no very convenient way of calculating the function $\varepsilon(\lambda_i, T)$ of metals for practical radiation measurements and pyrometry. Hence in studying the spectral distribution of radiating power for the majority of real bodies, especially metals, the function $\varepsilon(\lambda_i, T)$ is usually determined by experiment.

Since the results of various authors given below and their values of radiating power have a purely experimental character, we have endeavored to give them in the form of graphs and tables. Some data on the radiating power of metals are given in an Appendix. Naturally, the accuracy of determining radiating powers to a large extent depends on the method and apparatus used by the particular investigators. In this chapter, however, we shall simply consider very briefly the principles of the various methods of determining $\varepsilon(\lambda, T)$, $\varepsilon_\Sigma(T)$, $\rho(\lambda, T)$, not troubling with the construction of the apparatus.

The main methods of determining radiating power are based on the laws of thermal radiation. The complexity of absolute measurements of the energy flows of quasimonochromatic and total radiation often compels contemporary workers to avoid and practically to reject determining the values of $\varepsilon(\lambda, T)$ and $\varepsilon_\Sigma(T)$ by absolute measurements of the energy of nonblack radiation. For determining radiating powers, the most widely used are the following three methods, the physical essence of which we must briefly discuss.*

The first method is based on determining the radiating powers by measuring the ratio of the intensities of the radiation flux from a black body and the surface being studied, both being at the same temperature. By this method we may determine:

The spectral radiating power

$$\varepsilon(\lambda, T) = \frac{b(\lambda_l, T_l)}{b_0(\lambda_l, T_l)}$$

The total radiating power

$$\varepsilon_\Sigma(T) = \frac{\int\limits_0^\infty \varepsilon(\lambda, T) \, b_0(\lambda, T) \, d\lambda}{\int\limits_0^\infty b_0(\lambda, T) \, d\lambda} .$$

In practice this method is realized by creating a section with black radiation on the surface of the radiating substance.

The arrangement is shown schematically in Fig. 44. Here P is the sample for which the radiating power of the surface is to be determined. A black-radiation cavity ABCD is created in this. By due thermostatting, the

---

*The method of "regular thermal conditions" is briefly described on p. 40. The calorimetric method of determining $\varepsilon_\Sigma(T)$ is not considered in this book.

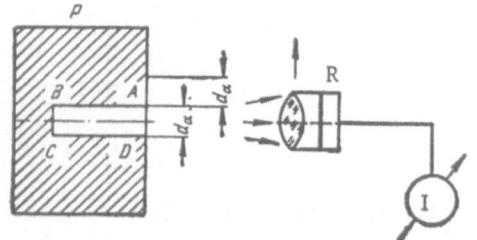

Fig. 44 Arrangement for determining radiating power from the radiation ratio.

temperature gradient on the surface and inside the cavity
is made to vanish.   The readings α and α_0 of indicator I
at the output of radiation receiver R are proportional to the
alternately incident radiation fluxes Φ and Φ_0 from the
surface $S_a = \pi d_a^2 / 4$ and the hollow ABCD with cross section
$S_a = \pi d^2/4$, respectively, in the spectral range defined by
the monochromatizing system M.   Then, clearly,

$$\Phi = S_a b(\lambda, T_x) \qquad \text{and} \qquad \Phi_0 = S_a b_0(\lambda, T_x).$$

Hence,

$$\varepsilon(\lambda, T_x) = \frac{\Phi}{\Phi_0} = \frac{\alpha}{\alpha_0}.$$

Using as radiation receiver R an ideal total-radiation receiver and accordingly removing the monochroma-
tizer M, we can determine the coefficient $\varepsilon_{\Sigma}(T_x)$. By measuring radiation fluxes at different angles in this
way, we can determine not only the normal but also the hemispherical values of radiating power. The main
advantage of this method is that the temperature does not have to be measured exactly; it is only important to
make sure that there is no temperature gradient between the surface and the hollow (we remember that, in accordance with Planck's law, the radiation flux depends very markedly on temperature). The need for creating
a black-radiation cavity in the substance being studied and for avoiding any temperature gradient between the
cavity and the surface imposes a serious limitation on the practicability of the method. These difficulties increase especially for materials with poor thermal conductivity; for metals, however, which have good thermal
conductivity and in which it is fairly easy to make the black-radiation cavity (because of their suitability for
mechanical working), this method finds wide application.

Naturally, the use of some kind of photometer system, measuring the brightness directly, in place of the
receiver R which records the energy of the incident radiation flux, simplifies the experimental conditions, since
it eliminates dependence on the area S_a and the distance. The values of spectral brightness as a function of
temperature, as determined by Planck's equation, are tabulated. To determine the spectral radiating power
in this case, it is convenient to use a brightness pyrometer, or still better a spectropyrometer.

A second method of determining spectral radiating power is to calculate it from measured values of
spectral brightness and true temperature. In the absence of a black-radiating cavity, this temperature is usually
determined by some kind of contact thermoelectric pyrometer. This method of determining $\varepsilon(\lambda, T)$ finds
practical application mainly in industrial pyrometry.

The third method is to determine the radiating power from the corresponding spectral or total reflection
coefficients. Thus for a nontransparent body we have from Kirchhoff's law:

$$\varepsilon(\lambda, T) = 1 - \rho(\lambda, T)$$

and correspondingly

$$e_{\Sigma}(T) = 1 - \rho_{\Sigma}(T),$$

where $\rho(\lambda, T)$ is the spectral reflecting power and $\rho_{\Sigma}(T)$, the albedo of the surface at temperature T.

Reflectometric methods of determining radiating powers in principle require neither the creation of a
black-body cavity nor a knowledge of the true temperature on the surface being studied. This constitutes a
great advantage of the reflection method. Determining the radiating power by reflectometric methods, however, is seriously complicated by the surface's own radiation and the difficulty of measuring reflected radiation when the reflection has a diffuse character. By using preliminary modulation of the incident radiation

flux from an auxiliary source, the effect of the self-radiation of the surface may be eliminated from the determination of reflecting power [29, 38].

Separation of the modulated flux of reflected radiation by resonance amplification of the photocurrents and synchronous detection at modulation frequency is in practice possible for any intensity of self-radiation from the surface under examination.

The modulation reflectometry of an incandescent surface such as that of a molten metal is illustrated schematically in Fig. 45.

Radiation is emitted by an auxiliary source GL (glow lamp) and is modulated by means of a rotating disk D_1: the movable mirrors M' and M" make it possible for this radiation, as reflected from the mirror RM, and the total radiation consisting of this radiation, as reflected from the surface S (e.g.,surface of molten metal) together with the thermal radiation of the surface itself, to pass alternately through the monochromatizing system MS into the receiver R.

The amplifier A, tuned to the modulation frequency, is furnished with a calibrated attenuator. The alternating component, consisting of pulses of sinusoidal voltage with amplitude proportional respectively to the incident and reflected radiation flux (depending on the positions of the mirrors M' and M") are measured by cathode voltmeters or recorded by the oscillograph I.

A total radiation flux Φ_T falls on the receiver; this consists of the steady flux of self-radiation from the incandescent or molten surface S, and an alternating flux Φ_A from the auxiliary source (for example, varying sinusoidally at modulation frequency ω) as reflected from the surface S, which has a reflecting coefficient of $\rho(\lambda, T)$.

Thus

$$\Phi_T = \Phi_0 + \Phi_A \rho(\lambda, T) \sin \omega t.$$

The resonance amplifier separates out only the component of frequency ω. Thus the recording system I measures $\alpha = \Phi_A(\lambda, T)$.* Clearly, for a known flux Φ_A the reflection coefficient $\rho(\lambda, T)$ is known, and for nontransparent bodies, by Kirchhoff's law, so is $\varepsilon(\lambda, T)$. Various forms of reflectometer are described in the literature [39-40].

In order to determine the coefficient of diffuse reflection, systems of the sphere type are used; these allow one to measure radiation reflected in all directions by the surface under examination. For high temperatures of the self-radiating surface, the difficulties of making absolute measurements of the flux reflected from the surface become matters of principle rather than technical matters. In recent years D. Ya. Svet [41] proposed a new method of determining the spectral reflection coefficients of a self-radiating surface, based on relative rather than absolute measurements of reflected fluxes. The essence of this method of modulation reflectometry, equally valid for mirror and diffuse reflection, is that the modulation reflectometer, in particular that operating on the principle shown in Fig. 45, determines not the absolute values of the spectral reflection coefficients at, for example, two wavelengths, $\rho(\lambda_1, T_i)$ and $\rho(\lambda_2, T_i)$, but their ratio

$$k = \frac{\rho(\lambda_1, T_i)}{\rho(\lambda_2, T_i)}. \tag{I}$$

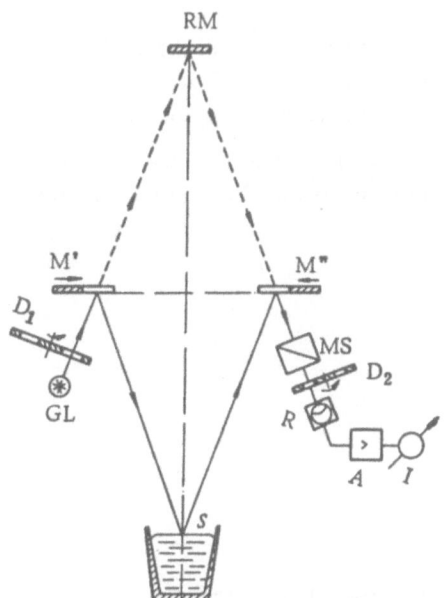

Fig. 45. Scheme of modulation reflectometer.

*Here as before we assume that the light characteristic of the receiver is linear, i.e., the reading is everywhere proportional to the flux.

Apart from this, the two spectral brightnesses $b(\lambda_1, T_i)$, $b(\lambda_2, T_i)$ of the self-radiation of the surface under examination are measured in the same direction by means of the same receiver R or an auxiliary receiver. But by definition

$$b(\lambda_1, T_i) = [1 - \rho(\lambda_1, T_i)]\, b_0(\lambda_1, T_i), \qquad \text{(II)}$$

$$b(\lambda_2, T_i) = [1 - \rho(\lambda_2, T_i)]\, b_0(\lambda_2, T_i). \qquad \text{(III)}$$

On simultaneous solution, the systems of equations (I), (II), and (III) enables us to calculate the three unknowns $\rho(\lambda_1, T_i)$, $\rho(\lambda_2, T_i)$ and T_i, the latter being the true temperature of the self-radiating surface. Naturally, this method allows us to do without an integrating sphere even for diffuse reflection. It should also be noted that, on using a third part of the spectrum with wavelength $b(\lambda_1, T_i)/b(\lambda_2, T_i)$ and $b(\lambda_2, T_i)/b(\lambda_3, T_i)$ and also correspondingly not one but two reflection-coefficient ratios: $\rho(\lambda_1, T_i)/\rho(\lambda_2, T_i)$ and $\rho(\lambda_2, T_i)/\rho(\lambda_3, T_i)$.

In this case the need to measure the absolute spectral brightnesses also disappears.

In order to determine the total radiating power, two methods are used: the calorimetric and "regular-condition" method, the former being an absolute-measurement approach. In this method the radiant flux is determined from the measured amount of heat directly given up by the body in question. With the calorimetric method it is also necessary to measure the radiant thermal fluxes.

With the method of regular thermal conditions it is unnecessary to measure the radiant thermal fluxes. This method does not demand the creation of a vacuum. The measuring apparatus and experimental technique are relatively simple. The experimental work on determining the emissivity reduces to determining the cooling rate. Samples of the solid under examination may be taken in any shape and of small sizes.

The method of regular conditions was used to determine emissivity by V. A. Osipova [42].

Here we consider the regular cooling of two bodies of arbitrary but similar geometrical size and shape in a constant-temperature medium, for one of which (the standard) the emissivity is known.

When the Biot criterion has a small value (≤ 0.1), the following relationships are valid for these:

$$\alpha = \frac{mGc_p}{S_1}, \quad \alpha_s = \frac{m_s G_s c_{ps}}{S_s},$$

where m and m_s are the cooling rates of the test and standard samples, G and G_s their masses, S and S_s their surface areas, c_p and c_{ps} the specific heats of the materials from which they are made, α_1 and α_2 the total heat-transfer coefficients, counting both convective and radiative heat exchange with the surrounding medium.

For identical thermal conditions of the experiment, the convective components of the heat-transfer coefficients for the bodies under comparison are the same, while the radiative components may be expressed by the well-known equations:

$$\alpha_r = \frac{c\left[\left(\dfrac{T_1}{100}\right)^4 - \left(\dfrac{T_2}{100}\right)^4\right]}{T_1 - T_2} = cf,$$

$$\alpha_{rs} = \frac{c_s\left[\left(\dfrac{T_1}{100}\right)^4 - \left(\dfrac{T_2}{100}\right)^4\right]}{T_1 - T_2} = c_s f_s.$$

Hence the difference in the radiant heat-transfer coefficients will be numerically equal to the difference in the total heat-transfer coefficients, which, allowing for the equations for α and α_s (with $S = S_s$) gives the relation

40

$$\Delta c = c_s - c = \frac{a_s - a}{f} = \frac{1}{fF}(G_s c_{ps} m_s - G c_p m).$$

Thus, we shall have the following computing equation for the emissivity:

$$\varepsilon_\Sigma = \frac{c_s - \Delta c}{c_0},$$

where c_0 is the radiation coefficient of an absolutely black body.

Measurement of Radiating Power in the Microwave Region

In the microwave region a method for determining the monochromatic radiating power based on classical radiotechnical measurement at high frequencies is used.

The essence of this method is given below, according to some somewhat improved* data of Vinokurov [43], using the very widely employed method of determining absorption losses (thermal losses in a dielectric) by the "tangent of the loss angle" (tg δ).

For dielectric materials one determines the emissivity in terms of the reflection coefficient of electromagnetic waves at the boundary between the media. One medium is taken as air with dielectric constant $E_1 = 1$ and tg $\delta_1 = 0$.

As second medium we have the substance under study, having dielectric constant E_2 and tg δ_2. The thickness of the dielectric is taken sufficient for complete absorption to take place within it.

The emissivity is determined in the direction normal to the surface of the dielectric.

For the above-mentioned conditions, the emissivity may be determined from the formula

$$\varepsilon(\lambda, T) = \frac{4\sqrt{2}\sqrt{E_2(1 + \sqrt{1 + \mathrm{tg}^2\delta_2})}}{\left[\sqrt{E_2(1 + \sqrt{1 + \mathrm{tg}^2\delta_2})} + \sqrt{2}\right]^2 + E_2(-1 + \sqrt{1 + \mathrm{tg}^2\delta_2})}.$$

For small tg $\delta_2 < 0.2$ and $E_2 > 1$ (case of real dielectrics), the formula may be considerably simplified:

$$\alpha_{f,T} = \frac{4\sqrt{E_2}}{(\sqrt{E_2} + 1)^2}.$$

The thickness of dielectric necessary for complete absorption is determined as follows.

The intensity and also the power of electromagnetic oscillations in a lossy dielectric vary with length according to

$$P_x = P_0 e^{-2\beta l},$$

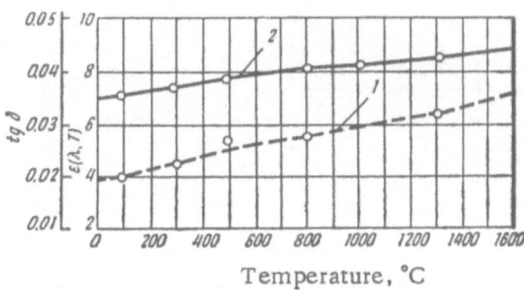

Fig. 46. Radiating power of polished corundum at $\lambda = 3$cm; sample 4 mm. 1) 1 tg δ; 2) $\varepsilon(\lambda, T)$.

*The improvement was kindly communicated to the author by G. L. Iossel'son, one of the leading specialists in the field of microwave pyrometry.

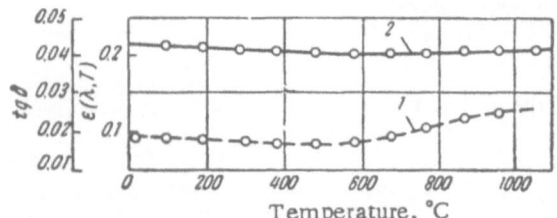

Fig. 47. Radiating power of polished corundum at $\lambda = 3$ cm; sample diameter 10 mm, placed perpendicular to the waveguide axis. 1) tg δ; 2) $\varepsilon(\lambda, T)$

whence

$$\lg \frac{P_x}{P_0} = -2\beta l \lg e.$$

Hence the attenuation per unit length in decibels will be (e = base of natural logarithms)

$$N = 10 \lg \frac{P_x}{P_0} = 20\beta \lg e \text{ dB/cm.}$$

where

$$\beta = \frac{2\pi}{\lambda} \sqrt{E} \text{ tg } \frac{\delta_2}{2},$$

or for $\lambda = 3.2$ cm

$$N = 17 \sqrt{E_2} \text{ tg } \frac{\delta_2}{2} \text{ dB/cm.}$$

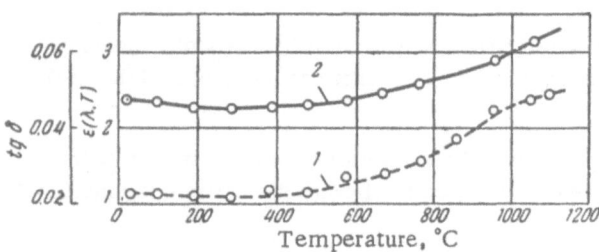

Fig. 48. Radiating power of corundum with mat surface at $\lambda = 3$ cm; sample diameter 10 mm. 1) tg δ; 2) $\varepsilon(\lambda, T)$.

Assuming the dielectric nontransparent with absorption 20 dB, the necessary thickness $l_{\min}$ is obtained in the form

$$l_{\min} = \frac{1.18}{\sqrt{E_2} \text{ tg } \frac{\delta_2}{2}}.$$

Investigations on the spectral radiating power of corundum (Al_2O_3) and silicon nitride (SiN) made by G. L. Iossel'son under the direction of V. V. Kandyba, showed that the results depended considerably on the emissivity of the material. A considerable part was also played by the orientation of the samples cut from the corundum single crystal when placed in the platinum waveguide in which measurements were made.

The curves of $\varepsilon(\lambda, T)$ and tg δ for two samples of polished corundum placed perpendicularly to the waveguide axis, plotted against temperature, appear in Figs. 46 and 47.

Correspondingly, data on $\varepsilon(\lambda, T) \times 10$ for a mat corundum surface are shown in Fig. 48; the radiating power $\varepsilon(\lambda, T) \times 10$ of silicon nitride is given in Fig. 49.

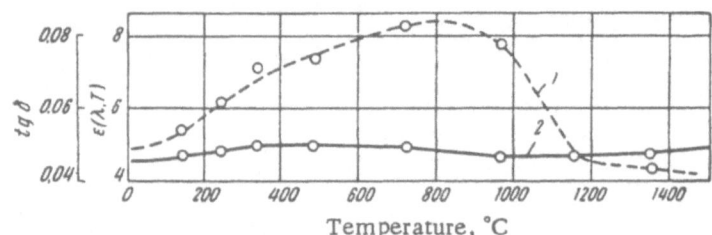

Fig. 49. Radiating power of silicon nitride for $\lambda = 3$ cm. 1) tg δ; 2) $\varepsilon(\lambda, T)$

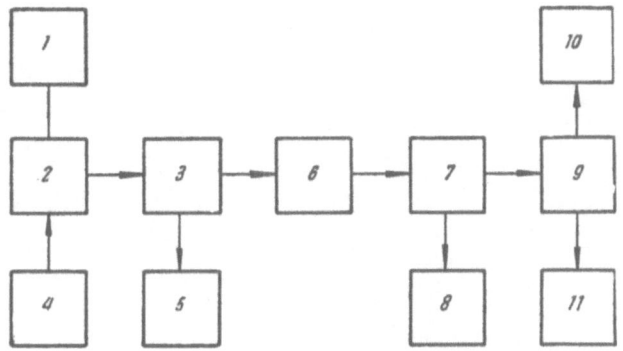

Fig. 50. Block diagram of $\varepsilon(\lambda, T)$ measurements in a short-circuited waveguide. 1) High-frequency generator $\lambda = 3$ cm); 2) ferrite modulator; 3) directed shunt (10 dB); 4) sound generator; 5) wavemeter; 6) attenuator; 7) measuring line; 8) measuring amplifier; 9) short-circuited waveguide; 10) millivoltmeter; 11) dc potentiometer.

On raising the temperature above a certain value, the losses in the sample rise very sharply. This agrees with the data of Bowie [44].

Kandyba and Iossel'son measured $\varepsilon(\lambda, T)$ from the value of tg δ at various temperatures, heating samples of the material under examination in a specially-constructed platinum short-circuited waveguide.

The block diagram of the experimental system of these authors is shown in Fig. 50.

CHAPTER IV

EXPERIMENTAL DATA ON THE RADIATING POWER OF METALS

The author's aim in this chapter is to acquaint readers with the results of investigations on radiating power conducted by a number of scientists over half a century. Naturally preference has been given to the best and most reliable examples of modern work, using the newest apparatus and most highly perfected methods. Except for two precision metrological studies on the radiating power of tungsten, the error in determining $\varepsilon(\lambda, T)$ absolutely is normally not smaller than 2 to 5%; this is because of the great dependence of radiating power on "surface purity" and also because of the difficulties associated with obtaining high accuracy in measuring radiant flux at high temperatures.

In individual cases of the radiating power of metals at high temperatures, including metals in the molten state, the measuring errors may be still higher. Additional reasons for reduced accuracy in determining the total radiating power $\varepsilon_{\Sigma}(\lambda, T)$ are the difficulties of eliminating chromatic aberration over a wide spectral range, especially when using radiation receivers with refracting ("lens") optical systems, and also the uncontrollable deviations of the radiation of the investigated surface from Lambert's law in the long-wave part of the spectrum, etc. Hence, while using the spectral and total radiating-power data presented as constants (for converting to true temperatures or in calculations), we have everywhere indicated the literary source which will enable the reader to acquaint himself most fully with specific methods and conditions capable of reflecting materially on the values of radiating power. Only isolated information is available on the radiating power of many metals, while for some there is none at all. The temperature affects the qualitative character of the relation between radiating power and wavelength comparatively little for most metals. Also useful are data on the spectral reflection coefficients of various metals at room temperature.

Data on the radiating powers of various metals are given in order of descending melting points.

Tungsten

This metal, which is the main emitter in incandescent lamps, plays a fundamental part in pyrometry. At the present time, with rare exceptions, the standard radiators and sources of radiation built into pyrometric systems constitute electrovacuum components with an incandescent part of tungsten (ribbon, "flat filament," etc.). The high melting point makes it possible to use incandescent tungsten lamps as standard sources of brightness temperatures up to 2500°C and color temperatures up to 2800°C.

Some quite extensive modern investigations of the spectral radiating power of tungsten were carried out in 1954 by De Vos [45] for the spectral range 0.25 to 2.6 μ. The results of this work for temperatures 1600 to 2800°K are presented in Fig. 51 on a semilogarithmic scale. The variable height (thickness) of the black band running along the axis of abscissas gives a quantitative indication of the absolute error in determining $\varepsilon(\lambda, T)$ for the corresponding wavelength. We see that the error increases on passing out of the visible region into the ultraviolet, and more especially into the infrared. This work was executed on a specially prepared radiator by comparing the radiation from the outer surface of a tungsten cylinder with that of the black body formed by the inside of the same cylinder. In 1959 Larrabee [46], working with a still more perfect radiator, showed that De Vos had failed to allow for error arising from additional radiation at the ends of the aperture, which distorted the black-body radiation. The extremely careful work of Larrabee embraced smaller temperature and wavelength ranges than that of De Vos: 1600 to 2400°K and 0.31 to 0.80 μ. The spectral radiating power of tungsten $\varepsilon(\lambda, T)$ is shown in Table 3.

Larrabee's [46] value of mean square error for the measurements was 0.2%. The total amount of impurity in the very pure tungsten (99.99%) used varied from 0.18 to 0.0018%. Figure 52 clearly shows the difference

TABLE 3

Wavelength λ, μ	Temperature, °K				
	1600	1800	2000	2200	2400
300	. . .	. . .	. . .	. . .	. . .
310	0.479	0.476	0.474	0.471	0.468
320	0.472	0.479	0.476	0.473	0.471
330	0.482	0.480	0.477	0.474	0.472
340	0.482	0.479	0.477	0.474	0.472
350	0.481	0.479	0.476	0.474	0.472
360	0.480	0.478	0.475	0.473	0.471
370	0.479	0.476	0.474	0.472	0.470
380	0.477	0.475	0.473	0.471	0.469
390	0.475	0.473	0.471	0.469	0.467
400	0.473	0.471	0.469	0.468	0.466
420	0.469	0.467	0.466	0.464	0.463
440	0.465	0.463	0.462	0.461	0.459
460	0.462	0.460	0.459	0.457	0.456
480	0.459	0.457	0.456	0.454	0.452
500	0.457	0.455	0.453	0.451	0.449
520	0.455	0.453	0.450	0.448	0.446
540	0.453	0.451	0.448	0.446	0.443
560	0.452	0.449	0.446	0.443	0.441
580	0.450	0.447	0.443	0.440	0.437
600	0.447	0.444	0.440	0.437	0.434
620	0.445	0.441	0.437	0.433	0.430
640	0.442	0.438	0.434	0.430	0.426
660	0.441	0.436	0.432	0.428	0.424
680	0.440	0.435	0.430	0.426	0.421
700	0.437	0.433	0.428	0.424	0.419
720	0.434	0.429	0.425	0.421	0.417
740	0.430	0.426	0.422	0.419	0.415
760	0.427	0.423	0.420	0.416	0.413
780	0.424	0.422	0.418	0.415	0.412
800	0.422	0.419	0.416	0.413	0.411

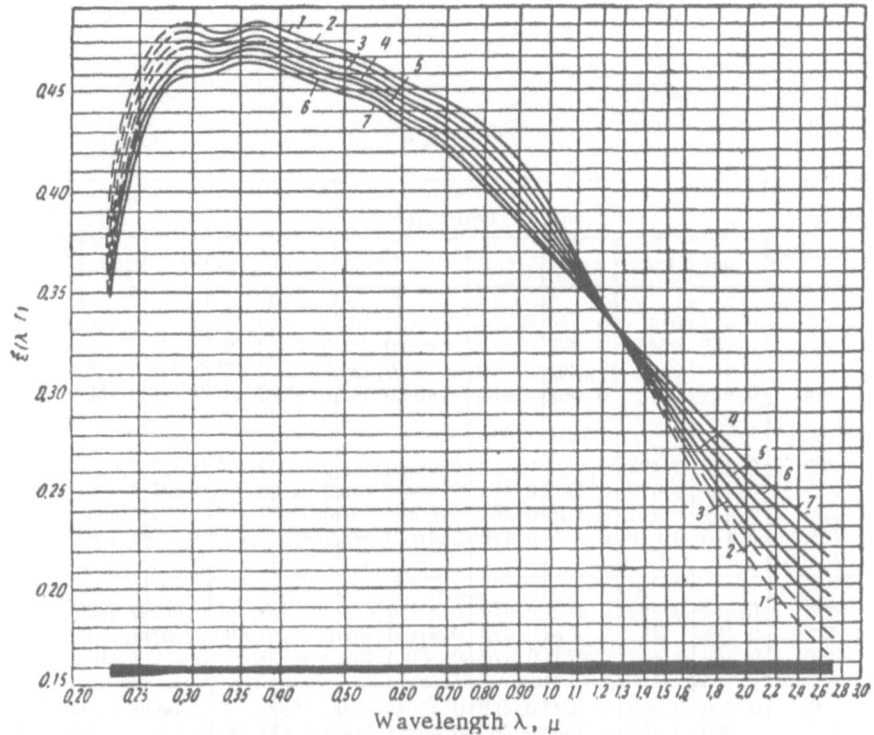

Fig. 51. Spectral radiating power of tungsten (De Vos). 1) 1600°K; 2) 1800°K; 3) 2000°K; 4) 2200°K; 5) 2400°K; 6) 2600°K; 7) 2800°K.

between the results of De Vos (broken lines) and Larrabee (continuous lines). In the visible part of the spectrum, especially in the green λ = 0.5 μ), the difference reaches 2% of the measured values of $\varepsilon(\lambda, T)$.

For determining the spectral radiating power in individual parts of the spectrum on a linear approximation, Larrabee derived the following empirical formulas:

For 350 to 450 mμ:

$$\varepsilon(\lambda, T) = 0.6075 - 0.3000\lambda - 0.3265 \cdot 10^{-4}\, T + 0.5900 \cdot 10^{-4}\,\lambda T$$

For 450 to 580 mμ

$$\varepsilon(\lambda, T) = 0.4655 + 0.01558\lambda + 0.2675 \cdot 10^{-4}\, T - 0.7305 \cdot 10^{-4}\lambda T$$

For 680 to 800 mμ

$$\varepsilon(\lambda, T) = 0.6552 - 0.2633\lambda - 0.733 \cdot 10^{-4}\, T + 0.7417 \cdot 10^{-4}\,\lambda T.$$

Larrabee's expressions [46] enable the value of $\varepsilon(\lambda, T)$ to be calculated with an error not exceeding 2.5%.

In view of the practical interest in the $\varepsilon(\lambda, T)$ of tungsten over a wider temperature range, Table 4 presents some older data (from [47] up to 1050 mμ and from Ornstein [48] above 1100 mμ; we must point out that the errors in $\varepsilon(\lambda, T)$ from these sources are considerably higher than those of Larrabee and De Vos). Very detailed, but unfortunately also fairly old, data are given by Ribault [9].

From these results, we present data on the total radiating power only of tungsten in Table 5.

TABLE 4

Wave-length λ, μ	Temperature, °K										
	1000	1200	1400	1600	1800	2000	2200	2400	2600	2800	3000
230	0.414	0.411	0.408	0.406	0.403	0.400	0.398	0.395	0.392	0.390	0.387
240	0.440	0.438	0.435	0.432	0.429	0.427	0.424	0.421	0.418	0.416	0.413
250	0.470	0.467	0.465	0.462	0.459	0.456	0.453	0.450	0.448	0.445	0.442
275	0.494	0.491	0.488	0.485	0.482	0.478	0.475	0.472	0.469	0.466	0.463
300	0.496	0.493	0.491	0.488	0.486	0.483	0.481	0.478	0.475	0.473	0.470
325	0.481	0.479	0.477	0.476	0.474	0.472	0.470	0.469	0.467	0.465	0.463
350	0.472	0.471	0.470	0.469	0.468	0.467	0.466	0.465	0.464	0.463	0.462
375	0.479	0.478	0.477	0.476	0.475	0.473	0.472	0.471	0.470	0.469	0.468
400	0.487	0.484	0.482	0.479	0.476	0.474	0.471	0.468	0.466	0.463	0.460
425	0.484	0.481	0.477	0.473	0.470	0.467	0.463	0.460	0.457	0.453	0.450
450	0.480	0.477	0.473	0.470	0.466	0.463	0.459	0.456	0.452	0.449	0.445
500	0.468	0.465	0.462	0.459	0.456	0.453	0.450	0.447	0.444	0.441	0.438
550	0.460	0.458	0.456	0.453	0.451	0.448	0.446	0.443	0.441	0.439	0.436
600	0.453	0.451	0.449	0.447	0.445	0.443	0.441	0.438	0.436	0.434	0.432
650	0.446	0.444	0.442	0.440	0.438	0.436	0.434	0.432	0.430	0.428	0.426
700	0.442	0.440	0.438	0.436	0.434	0.431	0.429	0.427	0.425	0.423	0.420
750	0.443	0.438	0.434	0.430	0.426	0.422	0.418	0.414	0.410	0.405	0.401
800	0.441	0.434	0.427	0.420	0.412	0.405	0.398	0.391	0.384	0.377	0.370
850	0.436	0.428	0.419	0.410	0.401	0.393	0.384	0.375	0.366	0.358	0.349
900	0.426	0.416	0.407	0.397	0.388	0.378	0.369	0.359	0.350	0.340	0.331
950	0.415	0.406	0.397	0.388	0.379	0.370	0.361	0.352	0.343	0.334	0.325
1000	0.396	0.388	0.380	0.372	0.364	0.355	0.347	0.339	0.331	0.323	0.315
1050	0.375	0.368	0.361	0.354	0.348	0.341	0.334	0.327	0.320	0.313	0.306
1100	—	—	—	0.345	0.334	0.322	0.311	0.299	0.288	0.276	0.265
1200	—	—	—	0.317	0.308	0.298	0.289	0.279	0.270	0.260	0.251
1300	—	—	—	0.295	0.288	0.280	0.273	0.266	0.258	0.251	0.244
1400	—	—	—	0.278	0.273	0.268	0.263	0.258	0.253	0.247	0.242
1500	—	—	—	0.264	0.261	0.258	0.256	0.253	0.250	0.247	0.244
1600	—	—	—	0.252	0.251	0.251	0.250	0.249	0.249	0.248	0.247
1700	—	—	—	0.242	0.243	0.245	0.246	0.247	0.248	0.250	0.251
1800	—	—	—	0.233	0.236	0.240	0.243	0.246	0.249	0.253	0.256
1900	—	—	—	0.225	0.230	0.234	0.239	0.243	0.248	0.252	0.257
2000	—	—	—	0.217	0.223	0.228	0.234	0.239	0.245	0.251	0.256

The newest data on the $\varepsilon(\lambda, T)$ of tungsten were obtained by Riethof et al. [49] by the method of comparing with black radiation (Fig. 53). These data are less accurate than those of De Vos and Larrabee, but nevertheless cover a fairly wide spectral range. The values of normal total radiating power calculated from these data are given below for tungsten.

The total hemispherical radiating power was also studied by Kudkin et al. [50] after heating tungsten for 6 to 10 h at 2800°K. Data from this investigation are shown by curve 1 in Fig. 54, curve 2 reflecting the results of Forsythe [51].

The above data on tungsten were obtained for the normal spectral radiating power. For example, in Larrabee's work [46], the maximum angle did not exceed 5°. Curves giving the variation in radiating power of a tungsten surface for ordinary and polarized light as a function of emission angle are given in Fig. 14.

Relationships for the spectral radiating power of tungsten ($\lambda = 0.66\ \mu$), 1) polished and unoxidized, 2) "slightly" oxidized, and 3) "fully" oxidized, according to Spulle [52], are shown in Fig. 55. The continuous lines in this figure show the results of Worthing for polished tungsten. As we see, oxidation causes the relationship between radiation and emission angle to approach Lambert's law.

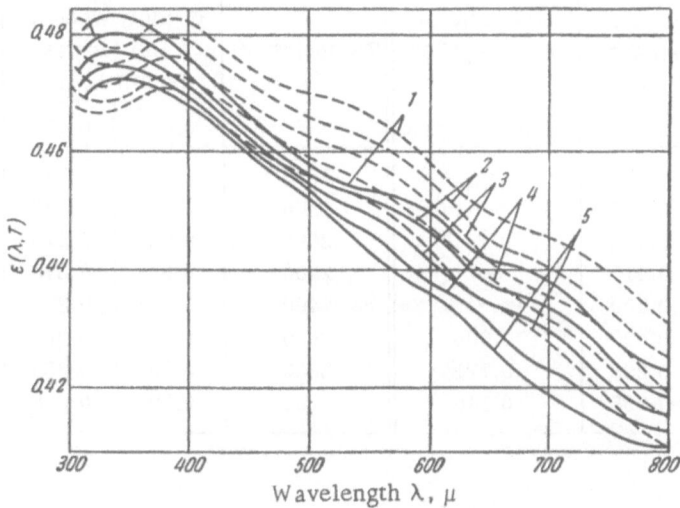

Fig. 52. Spectral radiating power of tungsten from data of Larrabee (continuous lines) and De Vos (broken lines). 1) 1600°K; 2) 1800°K; 3) 2000°K; 4) 2200°K; 5) 2400°K.

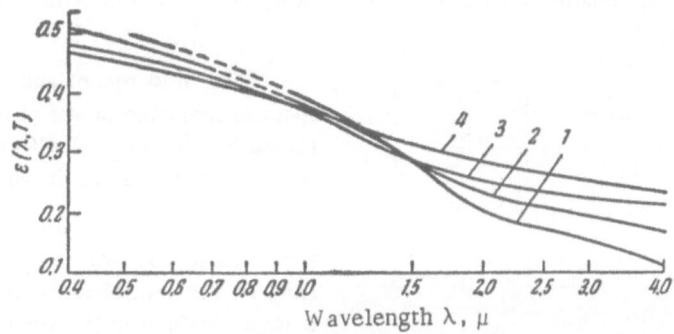

Fig. 53. Spectral radiating power of tungsten [49]. 1) 1600°K; 2) 2000°K; 3) 2800°K; 4) 3100°K.

TABLE 5

True temperature, °K	Total radiating power of tungsten $\varepsilon_\Sigma(\lambda, T)$	True temperature, °K	Total radiating power of tungsten $\varepsilon_\Sigma(\lambda, T)$	True temperature, °K	Total radiating power of tungsten $\varepsilon_\Sigma(\lambda, T)$
300	0.032	1400	0.175	2500	0.303
400	0.042	1500	0.192	2600	0.311
500	0.053	1600	0.207	2700	0.318
600	0.064	1700	0.222	2800	0.323
700	0.076	1800	0.236	3000	0.329
800	0.088	1900	0.249	3100	0.337
900	0.101	2000	0.260	3200	0.341
1000	0.114	2100	0.270	3300	0.344
1100	0.128	2200	0.279	3400	0.348
1200	0.143	2300	0.288	3500	0.351
1300	0.158	2400	0.296	3655	0.354

TABLE 6

Temperature, °K	Radiating power of molybdenum $\varepsilon(\lambda, T)$ for wavelength λ, μ		Total radiating power $\varepsilon_\Sigma(T)$	Temperature, °K	Radiating power of molybdenum $\varepsilon(\lambda, T)$ for wavelength λ, μ		Total radiating power $\varepsilon_\Sigma(T)$
	0.665	0.467			0.665	0.467	
273	0.420	0.425	—	1600	0.367	0.388	0.168
300	0.419	0.424	—	1800	0.360	0.383	0.189
400	0.415	0.421	—	2000	0.353	0.379	0.210
600	0.406	0.415	—	2200	0.347	0.375	0.230
800	0.398	0.409	—	2400	0.341	0.371	0.248
1000	0.390	0.403	0.096	2600	0.336	0.368	0.265
1200	0.382	0.398	0.121	2800	0.331	0.365	0.281
1400	0.375	0.393	0.145	2895	0.328	0.368	0.290

Molybdenum

Quite detailed but old data on the radiating power of molybdenum for the visible region of the spectrum, obtained by Worthing [53], are shown in Table 6. The spectral reflecting power of molybdenum for the infrared region appears in Fig. 56. [54].

At 1226°C the spectral radiating power of molybdenum was studied by Price [22] by comparing the radiation from the surface of a molybdenum cylinder with that of a black body. The data of this work appear in Table 7 and as curve 1 in Fig. 57.

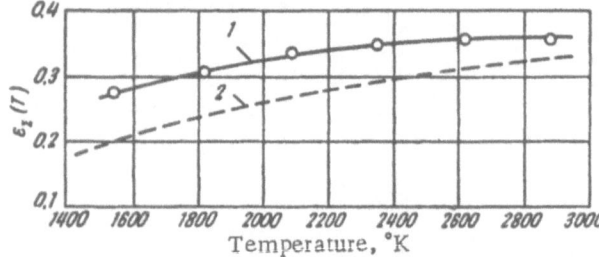

Fig. 54. Hemispherical total radiating power of tungsten as a function of temperature.

The most recent and complete results on the spectral radiating power of molybdenum were obtained by Riethof et al. [49] by comparing with a black body; these are shown as curves in Fig. 58.

As we see, the value of $\varepsilon(\lambda, T)$ with a zero temperature coefficient lies in the region $\lambda \approx 1.3$ μ. Data on the normal total radiating power of molybdenum obtained in the work in question by subsequent integration of the data on $\varepsilon(\lambda, T)$ are shown in Fig. 117 (curve 5).

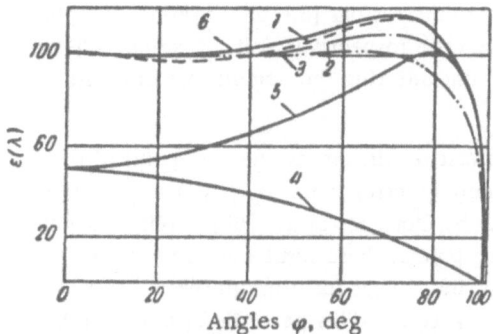

Fig. 55. Spectral radiating power of tungsten as a function of emission angle. 1) Surface polished and unoxidized; 2) "slightly" oxidized; 3) "fully" oxidized; 4) parallel-polarized component; 5) normally polarized component; 6) component of natural light (continuous curves correspond to data of Worthing [14]).

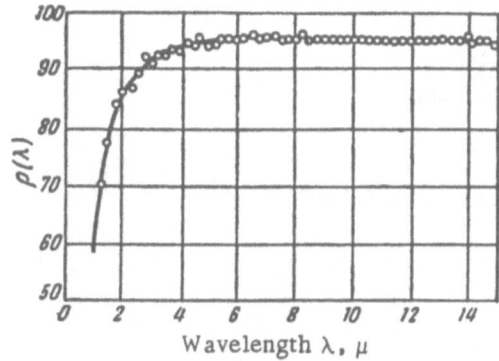

Fig. 56. Spectral reflecting power of molybdenum.

The hemispherical total radiating power of molybdenum [50] appears in Fig. 59 as triangles 2.

Tantalum

Data on the spectral radiating power of tantalum for two values of wavelength ($\lambda = 0.665\ \mu$ and $\lambda = 0.463\ \mu$) and the normal total radiating power [9] appear in Table 8. For $T = 1226°C$, data on the $\varepsilon(\lambda, T)$ of tantalum in the infrared region [22] are given in Fig. 57.

Up-to-date data on the spectral radiating power of tantalum in the visible and infrared region of the spectrum, obtained in the already-mentioned work of Riethof et al.[49], are shown in Fig. 60. The value of $\varepsilon(\lambda, T)$ with zero temperature coefficient occurs at $\lambda \approx 0.65\ \mu$. The still-not-published data of Marple (communicated by him to Riethof et al. [49]) are shown in the same figure by the broken line.

Rhenium

The results of individual determinations of the spectral radiating power of rhenium for the red part of the spectrum, carried out by various authors, are given by Levi [55] and Sims [56].

Detailed studies of the $\varepsilon(\lambda, T)$ of rhenium were made by Marple [57] in the spectral range of 0.35 to 2.8 μ for temperatures 1537 to 2772°C. Measurements were effected by comparing with a black-body radiation on a cylindrical rhenium cylinder. The construction of the radiator was as per De Vos (see above). Results on the spectral radiating power of rhenium for three temperatures are given in Fig. 61. Curve 4 gives the relationship for $\varepsilon(\lambda, T)$ calculated on the basis of classical electrodynamics by the formula of Hagen and Rubens (see above). Graphs of the variation in the temperature coefficient of the spectral radiating power are given separately in Fig. 62. Curve 1 shows the temperature coefficient of the radiating power, K_T, calculated from measurements of $\varepsilon(\lambda, T)$, and curve 2, that calculated from the Hagen and Rubens formula. As for tungsten, the "Drude boundary" for rhenium as the range of wavelengths starting from ~ 5 μ. Later Marple introduced corrections into some of his values [58].

TABLE 7

Wavelength $\lambda,\ \mu$	Radiating power of molybdenum $\varepsilon(\lambda, T)$	Wavelength $\lambda,\ \mu$	Radiating power of molybdenum $\varepsilon(\lambda, T)$
1.0	0.35	2.5	0.17
1.1	0.33	2.75	0.16
1.2	0.31	3.0	0.15
1.3	0.29	3.25	0.14
1.4	0.27	3.5	0.135
1.5	0.25	3.75	0.13
1.75	0.22	4.0	0.12
2.0	0.20	4.25	—
		4.5	0.1

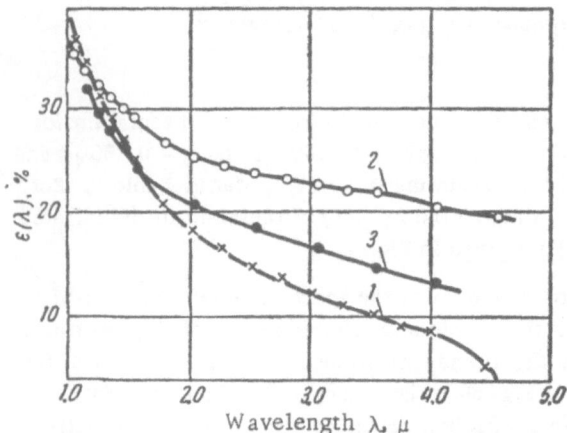

Fig. 57. Spectral radiating power of molybdenum at 1226°C. 1) Molybdenum, 2) iron, 3) nickel.

Osmium and Ruthenium

Recent investigations [59] on osmium and ruthenium have given the respective melting points as 3000 and 2250°C, contradicting the conclusions of earlier authors.

The newest results on the melting point of these metals and their spectral radiating power at $\lambda = 0.665 \, \mu$ were obtained by Douglass et al. [60]. With respect to the melting points, these results are very close to those of Baird [59]. Thus for osmium these temperatures are $2250 \pm 10°C$. This paper describes a method of determining $\varepsilon(\lambda, T)$ for the two materials using an optical pyrometer sighted first on the surface and then on a black-radiation cavity. The metals were of fairly high purity (up to $\sim 99.5\%$).

The author's values of spectral radiating power are shown as functions of temperature (curve 1 for osmium and 2 for ruthenium) in Fig. 63.

By analyzing the results, the authors derived the following formulas for calculating the $\varepsilon(\lambda, T)$ of the two metals at $\lambda = 0.665 \, \mu$:

for osmium

$$\lg_{10} \varepsilon(\lambda, T) = 9510 \frac{157.8 - 0.160T}{T(0.840T + 157.8)};$$

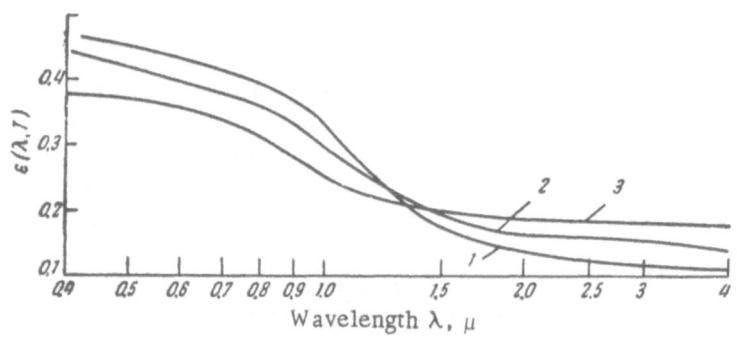

Fig. 58. Spectral radiating power of molybdenum. 1) At 1600°K; 2) at 2000°K; 3) at 2800°K.

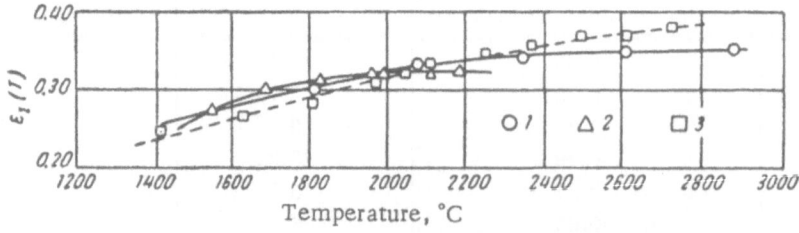

Fig. 59. Hemispherical total radiating power according to Kudkin at al. [50]. 1) Tungsten; 2) molybdenum; 3) rhenium.

TABLE 8

Temperature, °K	Radiating power of tantalum $\varepsilon(\lambda, T)$ wave-length λ, μ		Total radiating power $\varepsilon_\Sigma(T)$	Temperature, °K	Radiating power of tantalum $\varepsilon(\lambda, T)$ wave-length λ, μ		Total radiating power $\varepsilon_\Sigma(T)$
	0.665	0.463			0.665	0.463	
300	0.493	0.56	—	2200	0.411	0.46	0.251
1000	0.459	0.52	—	2400	0.404	0.45	0.669
1200	0.450	0.51	—	2600	0.397	0.44	0.287
1400	0.442	0.50	—	2800	0.390	—	0.304
1600	0.434	0.49	0.194	3000	0.384	—	—
1800	0.426	0.48	0.213	3300	0.375	—	—
2000	0.418	0.47	0.232				

for ruthenium

$$\lg_{10} \varepsilon\,(\lambda, T) = 9510 \,\frac{172{,}0 - 0{,}183T}{T\,(0{,}817T + 172{,}0)}.$$

A rather strange result which the authors were unable to explain satisfactorily is the existence of slight minima on the curves, at 2060°C for osmium and 1910°C for ruthenium.

For the majority of metals, in the visible region of the spectrum, the spectral radiating power diminishes with falling temperature. Thus, the existence of "minima" on the temperature curves of osmium and ruthenium force us to assume that, at any rate above 1900°C, the experiment or the measurements were methodically not without reproach. It should be noted that the assumption of a 1% temperature error in this investigation, which evidently took place in visual measurements with the brightness pyrometer, removes the perplexity regarding the "minima" arising in the $\varepsilon(\lambda, T)$.

Platinum

Various data on the spectral radiating power of platinum in the red part of the spectrum, assembled by Worthing [61], indicate the most reliable values as: for $\lambda = 0.658\,\mu$, those of Spence [62], given by curve 1 in Fig. 64; for $\lambda = 0.655\,\mu$, curve 2 [63] and curve 3 [64].

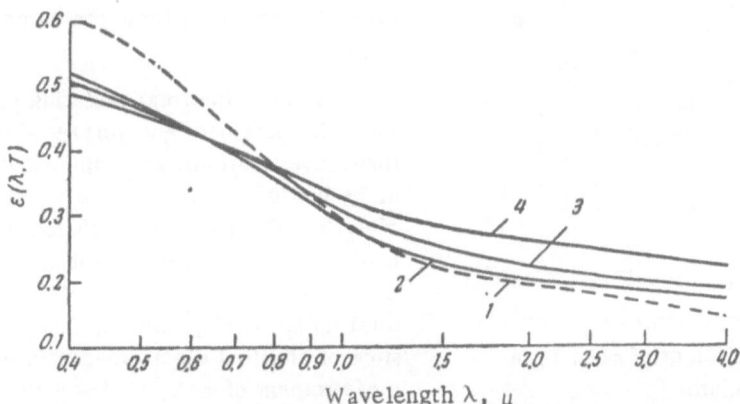

Fig. 60. Spectral radiating power of tantalum: 1) 1700°K; 2) 2200°K; 3) 2400°K; 4) 2800°K.

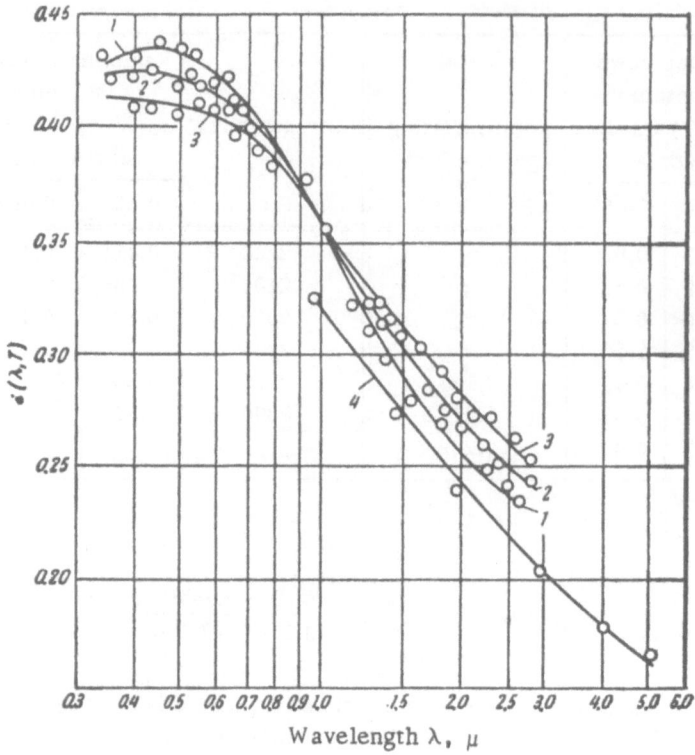

Fig. 61. Spectral radiating power of rhenium (lower curve 4 plotted by Drude's formula): 1) 1537°C; 2) 2000°C; 3) 2772°C.

The overestimated values of the $\varepsilon(\lambda, T)$ of platinum given in earlier papers are due to the fact that, apart from methodical errors, platinum of inadequate purity was used in the experiments. In the results given in Fig. 64, we notice an anomaly in the temperature variation of $\varepsilon(\lambda, T)$ for platinum. In contrast to other metals (see above), the $\varepsilon(\lambda, T)$ of platinum in the visible spectrum has a positive temperature coefficient. This might indicate that for platinum the point with the zero temperature coefficient of radiating power lay not in the infrared (as for tungsten, rhenium, etc.), but in the visible region. In order to resolve these doubts, discussed by Ribault more than thirty years ago, the $\varepsilon(\lambda, T)$ of platinum must be studied very carefully in the visible and infrared parts of the spectrum at different temperatures. For one value of temperature (T = 1125°C), the radiating power of platinum was determined by Price [22]. The results of these measurements appear in Table 9.

Data for the total radiating power of platinum were also obtained by several workers. The results of these investigations, according to Worthing [61], are given in Fig. 65. Here curve 1 corresponds to values of $\varepsilon_\Sigma(T)$ calculated by Foot's formula. This formula is of the two-term type and is analogous to that of Drude. A more precise formula for the hemispherical total radiation of platinum, also based on the conclusions of classical electromagnetic theory and data on the measurement of $\varepsilon(\lambda, T)$, takes the form

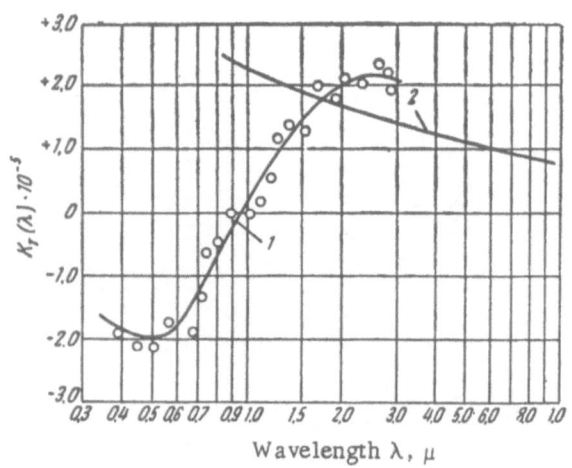

Fig. 62. Temperature coefficient of $\varepsilon(\lambda, T)$ of rhenium.

$$\varepsilon_\Sigma(T) = 0.892\,\Omega^{-1/3}\left(\frac{Tr_0}{c_2}\right)^{1/3} - 0.9047\,\Omega^{-1}\left(\frac{Tr_0}{c_2}\right) + 1.149\,\Omega^{-5/3}\left(\frac{Tr_0}{c_2}\right)^{5/3} - 1.245\,\Omega^{-2}\left(\frac{T}{c_2}\right)^2,$$

where r_0 is the specific electrical resistance of platinum.

A graph of $\varepsilon_\Sigma(T)$ plotted from this formula appears in Fig. 65, curve 4. Curves 2, 3, 5, and 6 in the same figure correspond to experimental measurements of the $\varepsilon_\Sigma(T)$ of platinum obtained by various workers. It should be noted that the data presented were calculated from the accepted value of $c_2 = 1.432\,\mu\cdot\deg$. The nature of angular dependence of the total radiating power and the spectral radiating power in the red part of the spectrum may be seen in Fig. 16.

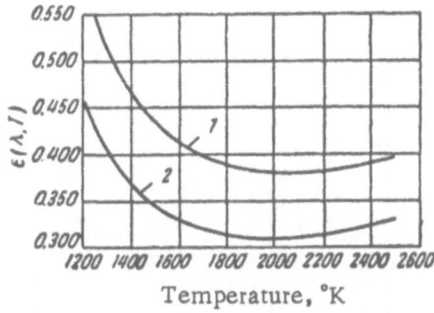

Fig. 63. Radiating power of osmium and ruthenium (at $\lambda = 0.665\,\mu$) as a function of temperature.

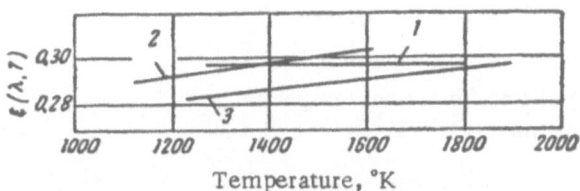

Fig. 64. Normal spectral radiating power of platinum as a function of temperature in the red part of the spectrum.

The most up-to-date data on the total hemispherical radiating power of polished platinum, due to Butler and Edward [25], are shown in Fig. 66.

Iron

Many investigations have been devoted to the radiating power of iron. Composite results on the spectral radiating power of iron for various temperatures from room temperature to the melting point are shown in Fig. 67. Here the open circles relate to the radiating power of iron calculated from the spectral reflection coefficients tabulated by Börnstein and Landolt [65]. The "pulses" and black rhombs correspond to $\varepsilon(\lambda, T)$ values obtained by modulation reflectometry for an iron bath at the melting point [66]. The values of $\varepsilon(\lambda, T)$ obtained at the same temperature for $\lambda = 0.65\,\mu$ by comparison with black-body radiation [67] are indicated by "x" signs. The values of $\varepsilon(\lambda, T)$ obtained at 1245°C [22] are shown by black circles. In Fig. 57 these data are presented in the form of curve 2, obtained by Price for the infrared part of the spectrum.

The variation of the $\varepsilon(\lambda, T)$ of iron in the solid state with temperature in the near infrared was also studied by Ward [68]. For a temperature of 1200°C the data from this work are shown by black triangles in Fig. 67.

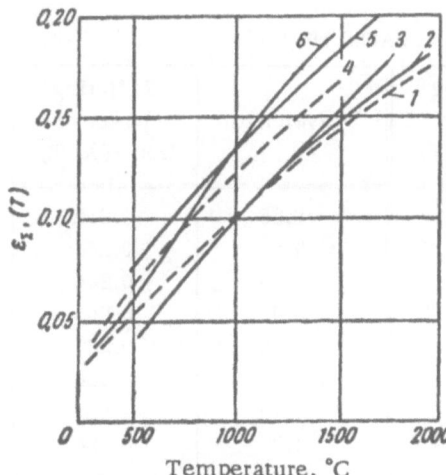

Fig. 65. Data of various workers in the total radiating power of platinum (broken curves theoretical).

TABLE 9

Wavelength λ, μ	Radiating power of platinum $\varepsilon(\lambda, T)$ at 1125°C	Wavelength λ, μ	Radiating power of platinum $\varepsilon(\lambda, T)$ at 1125°C
0.65	0.22	2.5	0.22
1.0	0.29	2.75	0.21
1.1	0.287	3.0	0.20
1.2	0.284	3.25	0.19
1.3	0.284	3.5	0.18
1.4	0.276	3.75	0.17
1.5	0.27	4.0	0.165
1.75	0.255	4.25	0.16
2.0	0.24	4.5	0.15
2.25	0.23	4.75	0.15

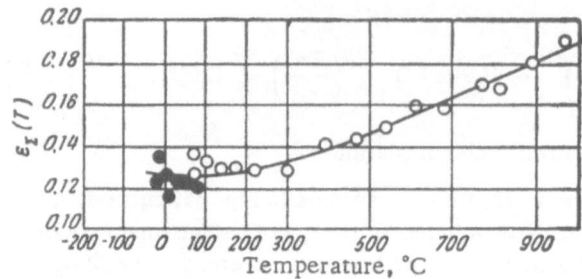

Fig. 66. Total hemispherical radiating power of platinum (black circles correspond to values at temperatures under 0°C).

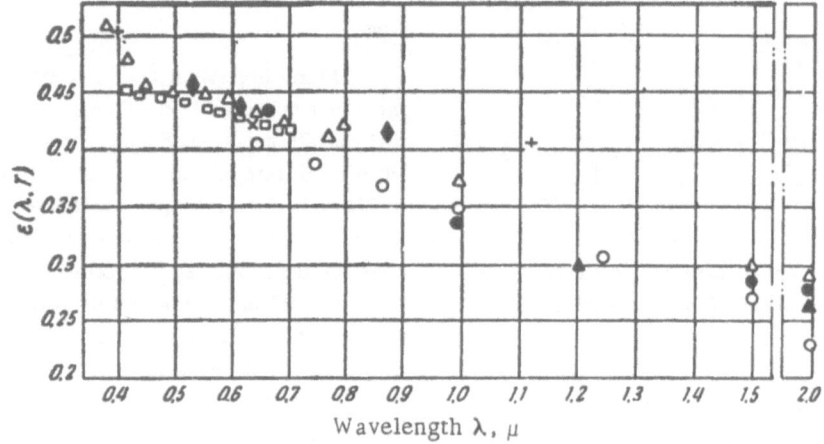

Fig. 67. Composite results on the $\varepsilon(\lambda, T)$ of iron for various temperatures obtained by different workers.

The data of Price on the spectral radiating power of iron at 1245°C are shown in Table 10.

Results on the temperature dependence of the spectral radiating power of iron in the near infrared, taken from the above-mentioned work by Ward [68], are shown as a family of curves in Fig. 68. The graphs

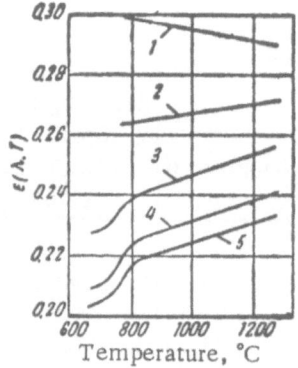

Fig. 68. Variation of the $\varepsilon(\lambda, T)$ of iron with temperature. 1) $\lambda = 1.2\,\mu$; 2) $\lambda = 1.6\,\mu$; 3) $\lambda = 2.0\,\mu$; 4) $\lambda = 2.4\,\mu$; 5) $\lambda = 2.7\,\mu$.

TABLE 10

Wavelength λ, μ	Radiating power of iron $\varepsilon(\lambda, T)$	Wavelength λ, μ	Radiating power of iron $\varepsilon(\lambda, T)$
0.65	0.437	2.25	0.252
1.0	0.34	2.5	0.248
1.1	0.33	2.75	0.244
1.2	0.316	3.0	0.24
1.3	0.31	3.25	0.237
1.4	0.3	3.5	0.235
1.5	0.29	3.75	—
1.75	0.27	4.0	0.225
2.0	0.26	4.25	—
		4.5	0.218

TABLE 11

Wavelength λ, μ	$\alpha \cdot 10^5$ deg^{-1}	Wavelength λ, μ	$\alpha \cdot 10^5$ deg^{-1}	Wavelength λ, μ	$\alpha \cdot 10^5$ deg^{-1}
1.0	− 10	1.8	+ 11	2.4	+ 17
1.2	− 7	2.0	+ 12	2.7	+ 18
1.4	− 6	2.2	+ 14	2.9	+ 20
1.6	+ 6				

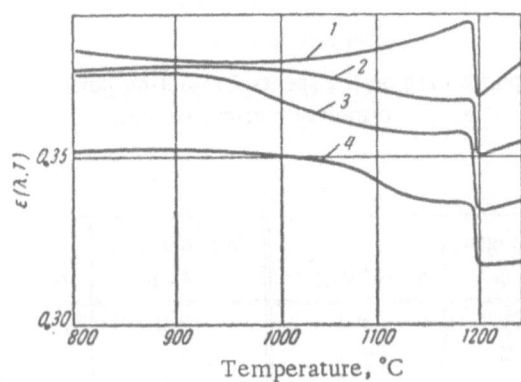

Fig. 69. Variation of the $\varepsilon(\lambda, T)$ of iron with temperature.
1) $\lambda = 0.45\ \mu$; 2) $\lambda = 0.55\ \mu$; 3) $\lambda = 0.65\ \mu$; 4) $\lambda = 0.85\ \mu$.

clearly show the bend in the $\varepsilon = f(\lambda)$ curve at a temperature corresponding to the Curie point. The values of the temperature coefficient for the spectral radiating power of iron calculated in this paper are given as $\alpha \cdot 10^5 \cdot$ deg^{-1} in Table 11.

The singular change in the $\varepsilon(\lambda, T)$ curve near the Curie temperature was also established earlier for the visible part of the spectrum [69]. The results of these experiments are shown in the form of a family of curves for the violet, green, red, and near-infrared parts of the spectrum in the temperature range 800 to 1200°K in Fig. 69.

The extremely serious technical difficulties connected with experiments to determine the radiating power of iron in the molten state compel us to treat the accuracy of the absolute values of $\varepsilon(\lambda, T)$ obtained by the radiation method with caution. In particular, the poor accuracy in determining $\varepsilon(\lambda, T)$ explains the contradictory data on the behavior of the temperature coefficient of the radiating power of a molten iron surface for the visible part of the spectrum, as given by various authors. Data on the radiating power of iron for the medium infrared region (up to 9 μ) at room temperature, according to Hase [70] and Coblentz [71], were presented as curves in Price's paper [72]. The same paper gives the $\varepsilon(\lambda, T)$ of iron for the near ultraviolet of room temperature.

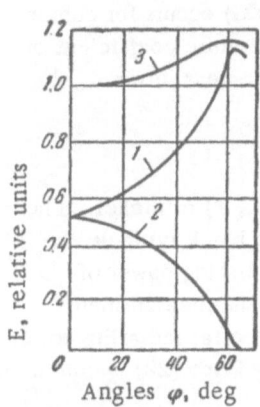

Fig. 70. Radiating power of iron as a function of emission angle.

Pepperhoff [73] studied the variation of the spectral radiating power of a molten-iron surface in the red part of the spectrum at $\lambda = 0.665\ \mu$ with angle of incidence for ordinary and polarized radiation. These results are shown as three curves in Fig. 70. Curves 1, 2, and 3 indicate, respectively, the parallel and normal polarized components and unpolarized light. Special interest attaches to the considerable increases in the $\varepsilon(\lambda, T)$ for the normal component (1) with increasing angle. For an angle of $\varphi = 80°$, the component of $\varepsilon(\lambda, T)$ reaches the radiation of a black body.

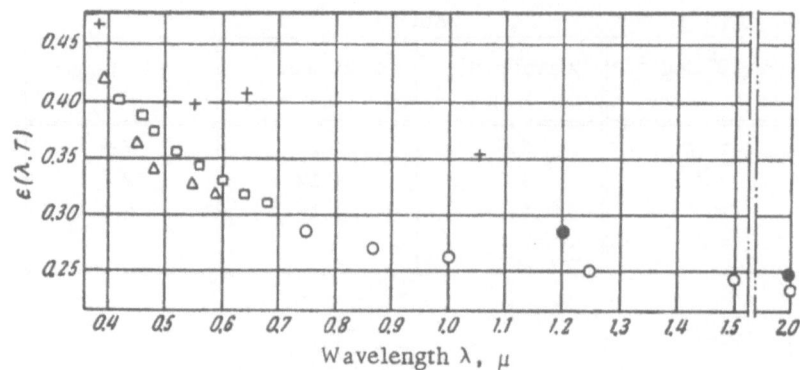

Fig. 71. Composite data on the spectral radiating power of cobalt obtained by various workers.

TABLE 12

Wavelength λ, μ	$\alpha \cdot 10^5$ deg^{-1}	Wavelength λ, μ	$\alpha \cdot 10^5$ deg^{-1}
1.1	− 8	2.2	+ 12
1.2	− 9	2.4	+ 15
1.4	− 5	2.6	+ 20
1.6	+ 3	2.8	+ 22
1.8	+ 6	3.0	+ 23
2.0	+ 8		

Cobalt

Composite data on the spectral radiating power of cobalt for various temperatures, from room temperature to the melting point, as given by various authors, appear in Fig. 71. The open circles give values of radiating power calculated from the spectral reflection coefficients obtained by various authors at room temperature, according to the tables of Börnstein and Landolt [65]. The "+" sign gives points for liquid cobalt at the melting point as found by modulation reflectometry [66]; the black circle indicates values obtained for $\varepsilon(\lambda, T)$ at 1000°C by Ward [68]. The latter determined the temperature dependence of $\varepsilon(\lambda, T)$ in the form of a family of curves (Fig. 72).

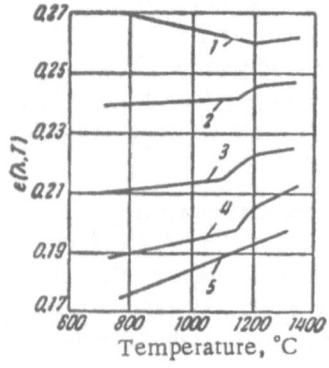

Fig. 72. Variation of the $\varepsilon(\lambda, T)$ of cobalt with temperature. 1) $\lambda =$ 1.2 μ; 2) $\lambda = 1.6$ μ; 3) $\lambda = 2.0$ μ; 4) $\lambda = 2.6$ μ; 5) $\lambda = 3.0$ μ.

Just as for iron, a marked bend representing a fairly sharp change in the radiating power near the Curie point (100°C) occurs for cobalt (also a ferromagnetic). Some data on the temperature coefficient of the spectral radiating power of solid cobalt in the near infrared, obtained by Ward [68] are shown in Table 12.

Nickel

Figure 73 shows composite data for the $\varepsilon(\lambda, T)$ of nickel. The notation is the same as for iron and cobalt. The black symbols, including the rhombs, give data on the spectral radiating power of nickel mirrors obtained by cathode sputtering. The measurements were made by various investigators [65]. The spectral radiating power of nickel in the solid state at 1110°C as given by Price [22] is indicated in Table 13 and Fig. 57.

The temperature dependence of the spectral radiating power of nickel in the near-infrared part of the spectrum (Ward [68]) is given Fig. 74.

58

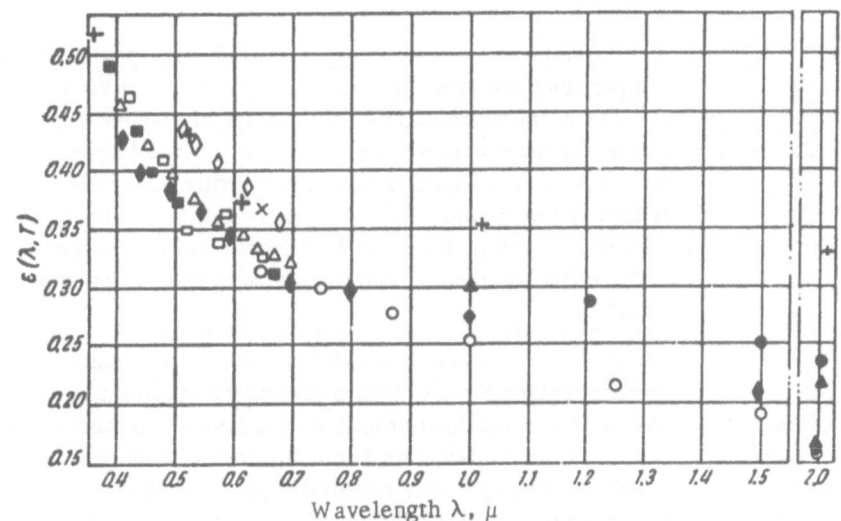

Fig. 73. Composite data on the $\varepsilon(\lambda, T)$ of nickel obtained by various authors.

TABLE 13

Wavelength λ, μ	Radiating power of nickel $\varepsilon(\lambda, T)$	Wavelength λ, μ	Radiating power of nickel $\varepsilon(\lambda, T)$
1.1	0.3	2.0	1.22
1.2	0.29	2.5	0.205
1.3	0.28	3.0	0.19
1.4	0.27	3.75	0.17
1.5	0.25	4.0	0.16

TABLE 14

Wavelength λ, μ	$\alpha \cdot 10^5$ deg^{-1}	Wavelength λ, μ	$\alpha \cdot 10^5$ deg^{-1}	Wavelength λ, μ	$\alpha \cdot 10^5$ deg^{-1}
1.0	− 10	1.8	+ 7	2.6	+ 13
1.2	− 5	2.0	+ 8	2.8	+ 18
1.4	− 3	2.2	+ 9	3.0	+ 19
1.6	+ 5	2.4	+ 11	3.2	+ 21

Values of the temperature coefficient of the spectral radiating power of nickel in the solid state (Ward) are given in Table 14.

Palladium

Data on the radiating power $\varepsilon(\lambda, T)$ of palladium are given in Table 15.

TABLE 15

State	Temperature, °K	Wavelength λ, μ	Radiating power of palladium $\varepsilon(\lambda, T)$	State	Temperature, °K	Wavelength λ, μ	Radiating power of palladium $\varepsilon(\lambda, T)$
Solid	1275	0.66	0.35	Solid	1805	0.55	0.33
''	1725	0.66	0.31			0.65	0.38
				Liquid	1830	0.65	0.37

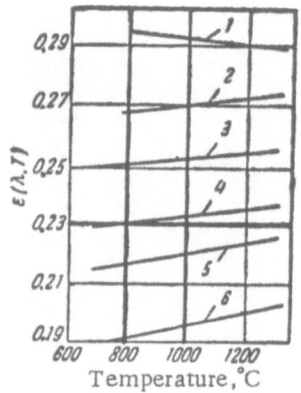

Fig. 74. Variation of the $\varepsilon(\lambda, T)$ of nickel with temperature. 1) $\lambda = 1.2\ \mu$; 2) $\lambda = 1.4\ \mu$; 3) $\lambda = 1.6\ \mu$; 4) $\lambda = 1.8\ \mu$; 5) $\lambda = 2.0\ \mu$; 6) $\lambda = 2.4\ \mu$.

Copper

Composite data on the spectral radiating power of copper at different temperatures are given in Fig. 75. Here the open circles show values of $\varepsilon(\lambda, T)$ at room temperature for mirrors of massive copper and the black circles for those obtained by sputtering. Just as before, the "+" indicates data for copper in the liquid state at the melting point, obtained by the modulation-reflectometer method. The "×" shows results of Stubbs [74] for molten copper, obtained by comparison with black-body radiation. The "□" indicates results of Price [22] for the radiating power of copper in the solid state at 900°C.

As we see, the optical anomaly of copper as compared with other metals (its yellow color in reflected light), and hence also its considerably smaller radiating power in this part of the spectrum as compared with the blue-green, is retained at high temperatures. In other words, the radiation of copper in the molten state is much richer in blue-green rays than red. This is indicated by the early fall in the $\varepsilon(\lambda, T)$ relationship in the visible part of the spectrum. Curve 1 in Fig. 76 was obtained by Stubbs for solid copper, and curve 2 for molten copper (visible part of the spectrum). These curves show that the change in $\varepsilon(\lambda, T)$ on increasing the wavelength in the visible part of the spectrum takes place more smoothly for molten than solid copper.

In the near infrared, however, the variation in the spectral radiating power of copper with falling wavelength is analogous to that of other metals. For illustration, Table 16 gives data on the $\varepsilon(\lambda, T)$ of copper obtained by Price [22].

The temperature dependence of the spectral radiating power of molten copper obtained by Stubbs [74] for two wavelengths is given in Table 17.

These data are fully confirmed in a paper by Vykhovskii [75], who measured the $\varepsilon(\lambda, T)$ with an optical pyrometer and a platinum−platinorhodium thermocouple.

Data on the spectral reflecting power of commercial copper in the infrared part of the spectrum (Gier and Dunkel [40]) are given in Fig. 77.

Values of $\varepsilon_{\Sigma}(T)$ for copper from the results of old investigations may be found in Worthing [61]. The total hemispherical radiating power of copper was shown in Fig. 28.

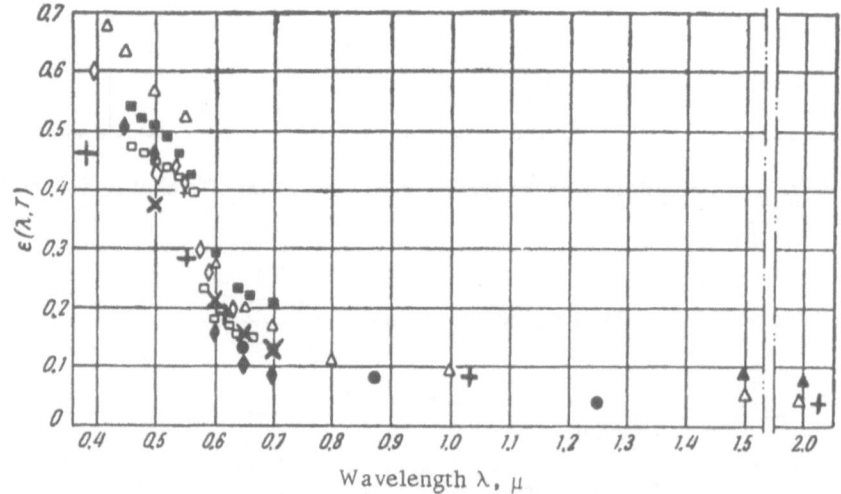

Fig. 75. Composite data on the $\varepsilon(\lambda, T)$ of copper obtained at various temperatures by various authors.

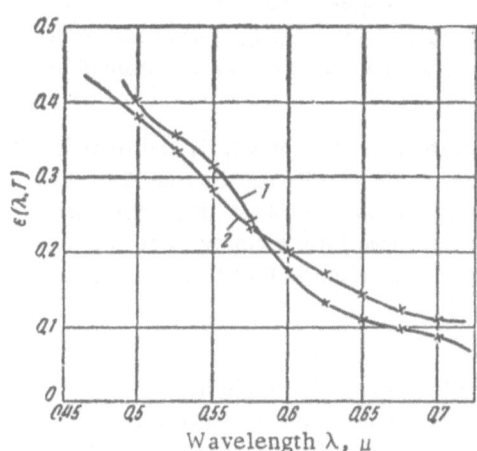

Fig. 76. Spectral radiating power. 1) Solid copper; 2) molten copper.

The closeness between the spectral radiating-power values of copper in the red part of the spectrum and the total radiating power, and also the above data on the radiating power of copper in the infrared, indicate that in the near-infrared region its radiation is approximately gray.

Gold

Values of the spectral radiating power of gold in the solid and liquid states obtained by Stubbs [76] are satisfactorily confirmed by later determinations of other authors (for example, composite data of Ribault [9]). We therefore restrict considerations to the $\varepsilon(\lambda, T)$ curves of Fig. 78 for gold in the solid and liquid states (Stubbs [76]).

The similarity between these curves and the corresponding data given above for copper is very striking. This becomes quite understandable when we remember the similarity in the "color" characterizing the spectral reflecting powers of copper and gold. Just as with copper, gold has a sharp region of anomalous dispersion, not in the ultraviolet, but in the blue-green part of the spectrum. As with copper, the collapse of crystal structure on melting makes the course of the $\varepsilon(\lambda, T)$ relationship for molten gold smoother. The reflecting power of gold obtained by electrolysis, as given by Gier et al. [40], is shown in Fig. 79 for the infrared part of the spectrum.

Silver

Composite data on the spectral radiating power of silver are given in Fig.80 (notation as in Fig. 79). It should be noted that with silver, which like all other metals save gold and copper has a "white color", the region of anomalous dispersion lies closer to the visible region than, for example, with iron. This is confirmed not only by the course of the $\varepsilon(\lambda, T)$ curve given for a cold surface, but also by certain results on the transmission of thin silver films [77]. These films have a violet color in transmitted light. The value $\varepsilon_\Sigma(T)$ for silver at 1000°C, according to Ribault [9], is 0.035. At room temperature, according to Wood [77], $\varepsilon_\Sigma(T) = 0.02$ for silver.

Aluminum

A detailed study of the spectral radiating power of aluminum in the infrared part of the spectrum was made by Hase [37] for temperatures up to 600°C. The values of $\varepsilon(\lambda, T)$ for a specially etched (curve 1) and simply polished (curve 2) aluminum surface obtained in this work and shown in Fig. 81 indicate that, beginning from approximately 2 μ, the radiation of aluminum is nearly gray, i.e., $\partial\varepsilon(\lambda, T)/\partial\lambda \approx 0$.

TABLE 16

Wavelength λ, μ	Radiating power of copper $\varepsilon(\lambda, T)$	Wavelength λ, μ	Radiating power of copper $\varepsilon(\lambda, T)$
1.5	0.8	3.0	0.4
2.0	0.7	3.5	0.38
2.5	0.5	4.0	0.3

Note. Experiments made at 900°C for the infrared part of the spectrum

TABLE 17

Wavelength λ, μ	Radiating power of molten copper $\varepsilon(\lambda, T)$ at temperature (°C)			
	1075	1125	1175	1125
0.65	0.17	0.15	0.14	0.13
0.55	0.47	0.38	0.32	0.28

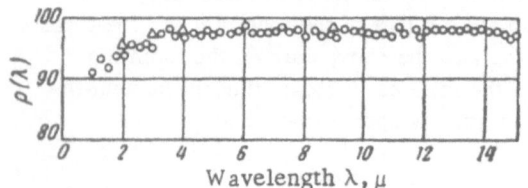

Fig. 77. Spectral radiating power of copper in the infrared part of the spectrum.

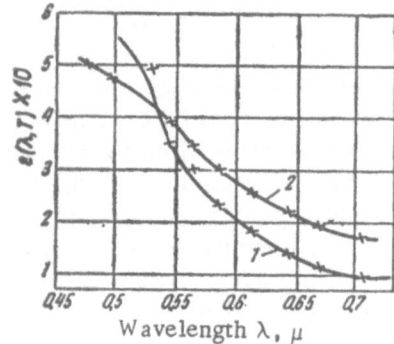

Fig. 78. Spectral radiating power: 1) Solid gold; 2) molten gold.

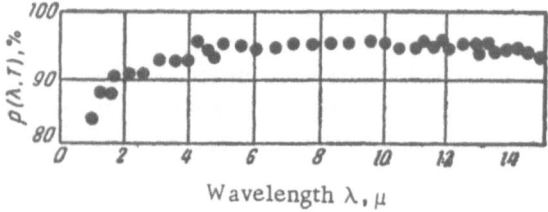

Fig. 79. Spectral reflecting power of gold in the infrared part of the spectrum.

The effect of the degree of surface roughness on the $\varepsilon(\lambda, T)$ of aluminum may be judged from the curves for polished, mat, and sand-blasted surfaces obtained at 400°C (Fig. 82).

In this figure the Roman figures I, II, and III indicate the spectral radiation-density distributions $E(\lambda, T)$ for the correspondingly treated surfaces. For comparison we also show here the curve for a black body at 400°C (indicated by "0"). The ordinates E of these curves are given in relative units.

Analogous results obtained by Hase for cast aluminum are shown in Fig. 83 and for Duralumin in Fig. 84. In all these figures the notation for polished, mat, and sand-blasted surfaces are as before (1, 2, and 3). The spectral radiating power of an aluminum polished surface also changes very little with temperature. This may be seen in Fig. 85. In view of the special interest which aluminum mirrors present in technology, Table 18 gives some figures for the spectral reflecting power of these [78]. The results relate to the reflecting powers of mirrors manufactured from "superpure" (99.99%) and "ordinary purity" aluminum. We see the effect of aging on the spectral reflecting power of both kinds of aluminum. We see the effect of aging on the spectral reflecting power of both kinds of aluminum; the value $\rho(\lambda)$ corresponds to data obtained immediately after depositing the mirror, and $\rho'(\lambda)$ to eight days later.

The reflecting power falls as a result of surface oxidation.

Data on the total normal reflecting power at 400°C for aluminum and Duralumin, given by Hase, are shown in Table 19.

From the polar diagrams of the spectral radiating power of aluminum and Duralumin surfaces as functions of emission angle given in this paper for 400°C and wavelength $\lambda = 3\,\mu$, we may conclude that, for a rough surface, the variation in radiation intensity with emission angle corresponds most nearly to a cosine law (Lambert's law).

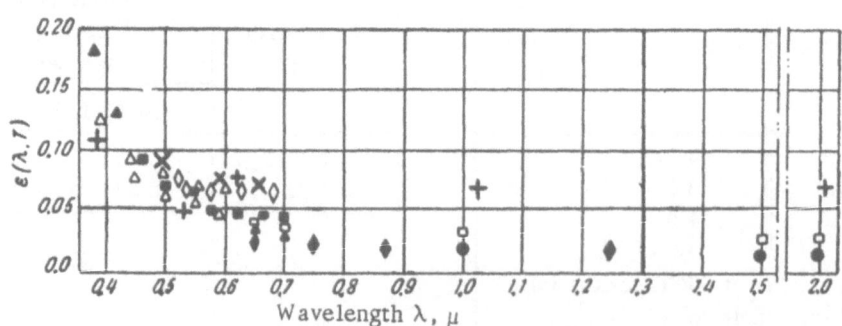

Fig. 80. Composite data on the spectral radiating power of silver.

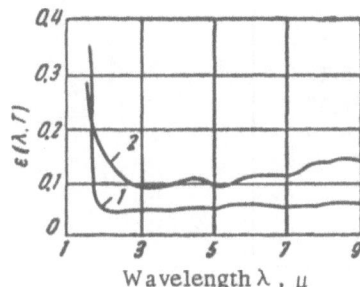

Fig. 81. Spectral radiating power of an aluminum mirror.

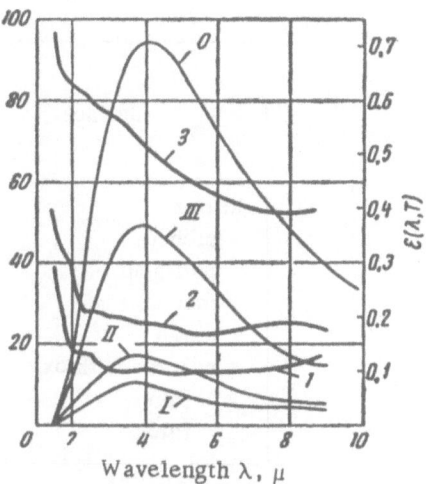

Fig. 82. Spectral radiating power of aluminum. 1) Polished; 2) mat; 3) after sand-blasting.

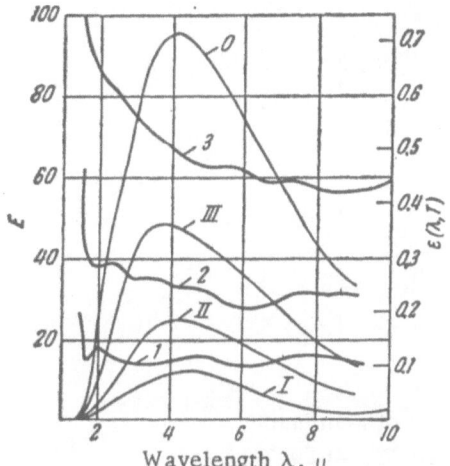

Fig. 83. Spectral radiating power of cast aluminum. 1) Polished; 2) mat; 3) after sand-blasting.

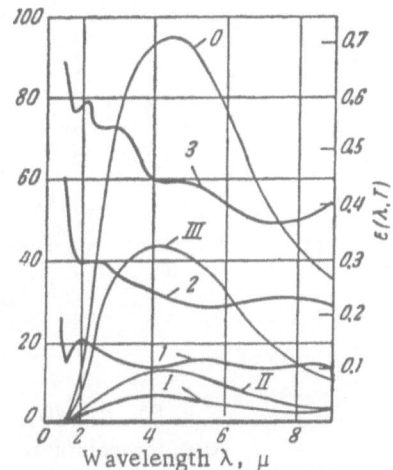

Fig. 84. Spectral radiating power of Duralumin. 1) Polished; 2) mat; 3) after sand-blasting.

Data on the relative intensity as a function of emission angle are presented in Table 20.

Nickel—Iron and Nickel—Copper Alloys

The spectral radiating power of binary alloys in the molten state was studied for wavelength $\lambda = 0.650\ \mu$ and temperature 1535°C in [79].

TABLE 18

Wavelength λ, μ	Aluminum 99.99%		Aluminum 99.99%		Wavelength λ, μ	Aluminum 99.99%		Aluminum 99.99%	
	$\rho(\lambda)$	$\rho'(\lambda)$	$\rho(\lambda)$	$\rho'(\lambda)$		$\rho(\lambda)$	$\rho'(\lambda)$	$\rho(\lambda)$	$\rho'(\lambda)$
0.46	0.92	0.895	0.89	0.86	0.57	0.93	0.885	0.90	0.85
0.53	0.92	0.895	0.89	0.86	0.60	0.93	0.880	0.90	0.86

TABLE 19

Surface	Radiating power		
	aluminum		Duralumin
	pure	cast	
Mirror			
etched	0.07	—	—
polished	0.11	0.13	0.12
mat	0.19	0.27	0.25
rough	0.475	0.49	0.47
covered with oxide film:			
thin	0.36	—	—
thick	0.45	—	—

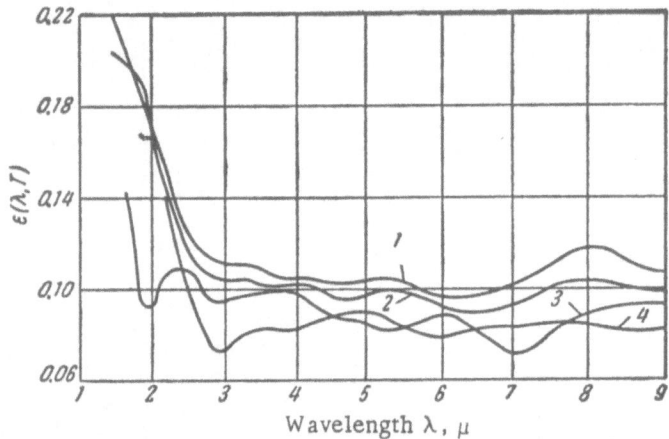

Fig. 85. Variation of spectral radiating power of aluminum with temperature. 1) 400°C; 2) 600°C; 3) 300°C; 4) 500°C.

TABLE 20

Surface	Radiating power at emission angle (deg):								
	0	10	20	30	40	50	60	70	80
Aluminum									
Polished	1.0	0.86	0.81	0.82	0.89	0.95	1.06	1.24	1.63
Mat	1.0	1.0	1.01	1.04	1.10	1.17	1.29	1.38	1.45
After sand-blasting	1.0	1.0	1.0	1.0	1.0	0.99	0.98	0.95	0.90
Slightly oxidized	1.0	0.86	0.79	0.76	0.79	0.82	0.91	0.99	1.01
Strongly oxidized	1.0	0 94	0.95	1.01	1.10	1.19	1.24	1.25	1.21
	1.0	0.85	0.84	0.88	0.99	1.13	1.36	1.63	1.88
Duralumin									
Polished	1.0	0.86	0.79	0.79	0.82	0.88	0.96	1.11	1.60
Mat	1.0	0.86	0.83	0.85	0.89	0.95	1.03	1.14	1.24

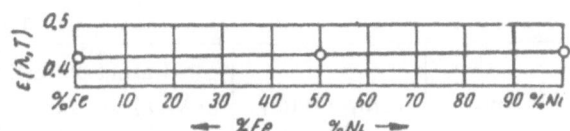

Fig. 87. $\varepsilon(\lambda, T)$ for λ = 0.650 μ for iron—nickel
alloys.

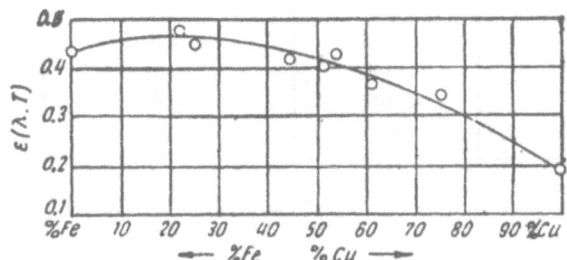

Fig. 86. $\varepsilon(\lambda, T)$ for λ = 0.650 μ for iron—copper
alloys

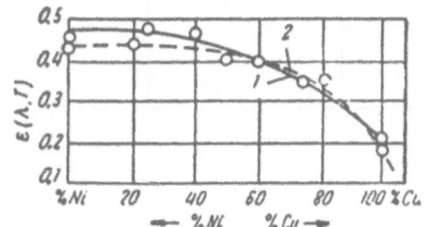

Fig. 88. $\varepsilon(\lambda, T)$ for nickel—copper alloys: 1) λ =
0.62 μ [80]; 2) λ = 0.65 μ [79].

Data for iron—copper and iron—nickel alloys appear, respectively, in Figs. 86 and 87. Figure 88 gives comparative data for nickel—copper alloys; the broken line is for λ = 0.65 μ (from [79]), and the continuous line for λ = 0.62 μ (from [80], using the modulation reflectometer).

As we see, the reproducibility of the results is extremely satisfactory.

The abscissas give the mass percentages of the alloy components. The spectral radiating power for the iron alloys was measured at 1535°C and for the nickel at 1450°C.

Figure 89 presents a family of $\varepsilon(\lambda, T)$ curves for nickel—copper alloy obtained in the modulation reflectometer [81]. Results of investigations on the temperature dependence of $\varepsilon(\lambda, T)$ for several alloys, especially stainless steel, titanium, nickel, Inconel, etc., may be found in articles by Poggie [82] and Beavens et al. [83]. Further information is in Bloch's monograph [84].

Germanium and Silicon

The spectral radiating power of germanium and silicon at wavelength λ = 0.65 μ was obtained by Allen [85] by comparing with the radiation of a black body by means of an optical micropyrometer. These results are listed for pure germanium and silicon surfaces, respectively, in Table 21 and illustrated in Fig. 90 as curves 1 and 2. The measurements were made in vacuum. The broken curve 3 corresponds to values of $\varepsilon(\lambda, T)$ for a silicon surface measured in air.

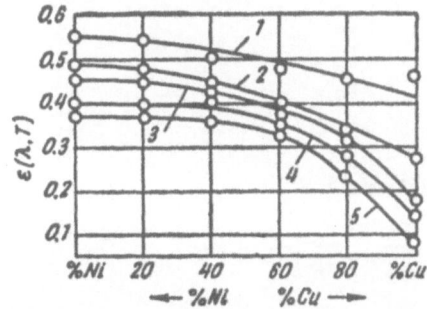

Fig. 89. Spectral radiating power of nickel—cop-
per alloys: 1) λ = 0.39 μ; 2) λ = 0.54 μ; 3) λ =
0.62 μ; 4) λ = 1.03 μ; 5) λ = 2.17 μ.

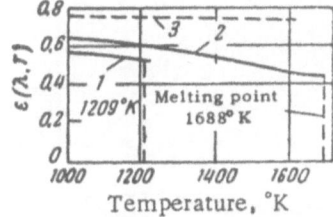

Fig. 90. Temperature variation of the
spectral radiating power of: 1) germanium; 2) silicon; 3) silicon (in air).

TABLE 21

Material	Temperature, °K	Radiating power $\varepsilon(\lambda, T)$ for wave-length $\lambda = 0.65\,\mu$	Material	Temperature, °K	Radiating power $\varepsilon(\lambda, T)$ for wave-length $\lambda = 0.65\,\mu$
Silicon	1000	0.64	Silicon	1500	0.50
	1100	0.62		1600	0.48
	1200	0.60		1688	0.46
	1300	0.57	Germanium	1000	0.56
	1400	0.54		1100	0.55
				1200	0.53

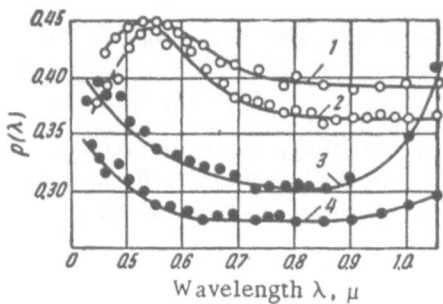

Fig. 91. Spectral reflecting power. 1), 2) Germanium; 3), 4) silicon.

The reflecting power of germanium and silicon was studied by Vavilov et al. [86] with a reflectometer incorporating an integrating sphere. The error of these measurements lay within ∼ 2%.

The results are shown in Fig. 91, where curve 1 corresponds to a polished n-type germanium single crystal (r ∼ 10 Ω · cm) not subjected to etching; curve 2 is for the same germanium samples after deep etching in H_2O_2; curve 3 is for a polished p-type silicon single crystal (r ∼10 Ω · cm), and curve 4 for a phosphorus-doped in n-type silicon single crystal (surface concentration of the order of 10^{17} cm^{-3}).

The data obtained indicate that the reflection coefficient depends substantially on the treatment of the surface.

Data on the value of $\varepsilon(\lambda, T)$ for germanium and indium antimonide in the near infrared were obtained by Moss and Hawkins [87].

CHAPTER V

EXPERIMENTAL DATA ON THE RADIATING POWER
OF CARBON AND GRAPHITE

The unique role played by graphite in modern high-temperature technology is generally known. The reflecting powers of pure, noncrystalline carbon and graphite have been the subject of many scientific studies. The spectral radiating power of graphite of wavelength 0.665 μ was initially studied as soon as optical pyrometers appeared.

Carbon and graphite have extremely diverse properties, due to their electron bonds, structures, etc. The radiating power of ordinary carbon and graphite, naturally, is a function of the treatment of the sample surface, which, as mentioned above, is extremely difficult to determine quantitatively. The spectral radiating power has been measured as a function of wavelength for noncrystalline carbon and graphite by various investigators.

The data given below are taken from the most modern and extremely circumstantial work of Plankett and Kingery [88]. The results shown in Fig. 92, except those of Euler [89] and MacPherson [90], which were obtained for high temperatures in an arc, represent values of reflection coefficients obtained by various workers [89] at room temperatures, for which the radiating powers are calculated by Kirchhoff's law. Despite the substantial differences in the values, we note the characteristic fall in the spectral radiating power for increasing wavelength, which is typical for metals.

The variegation of the data is explained by the different types of materials and the nature of the surface treatment. The results of Coblentz (curves 1 and 2 in Fig. 92) [92] illustrate the influence of the material and character of the surface. For a sample of "Siberian" graphite, carefully polished, the radiating power was considerably lower (curve 1) than for a piece of Acheson graphite, which, after polishing, remained fairly porous (radiating power shown in curve 2).

These curves indicate a fall in reflecting power due to energy scattering by rough and uneven parts of the surface in measuring the mirror-reflected ray.

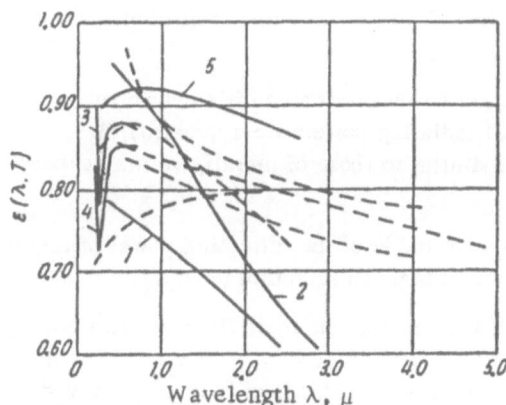

Fig. 92. $\varepsilon(\lambda, T)$ of graphite and carbon at various temperatures, according to the data of various workers.

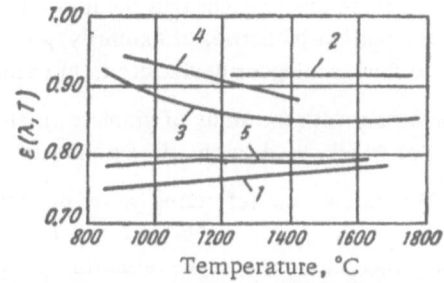

Fig. 93. Temperature variation of the $\varepsilon(\lambda, T)$ of graphite and carbon at $\lambda = 0.65\ \mu$, from the data of various workers.

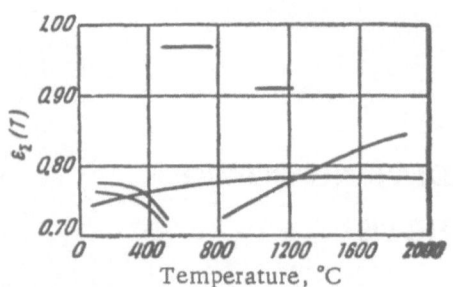

Fig. 94. Composite data on the temperature dependence of the $\varepsilon_\Sigma(T)$ of graphite and carbon as given in various papers.

Clearly, graphite cannot be regarded as a gray body when considering the infrared part of the spectrum. Unpolished graphite samples, however, may serve as a fairly good approximation to a gray body in the visible spectrum.

Humphrey-Owen and Gilbert [93] (curves 3 and 4 in Fig. 92) established the existence of a radiating-power minimum and a reflecting-power maximum at $\lambda = 0.260\ \mu$ for graphite of several types with different surface characters. Polished and ground surfaces have comparatively low radiating power. Curves 3 and 4 represent values obtained as a result of treating the surface of pure graphite samples in different ways. The existence of this spectral minimum is also observed in the spectrum of the carbon molecules for the aromatic series at the same wavelength. Measurements for noncrystalline carbon and for anthracite show that this minimum also exists here.

Measurements of the reflecting power of graphite made by Betz (curve 5) [94] also indicate the existence of this minimum. Unfortunately, the band was not extended to the shorter-wave side in this work. Various workers measured the spectral radiating power more precisely for the red part of the spectrum ($\lambda = 0.66\ \mu$) in the temperature range 800 to 2000°C, but the discrepancies between their respective results were often not only quantitative but also qualitative [different temperature coefficients of $\varepsilon(\lambda, T)$]. Figure 93 gives results of several investigations, in particular those of Thorn and Simpson, on the radiating power of carbon and graphite. We may well consider this work on account of its meticulous execution [95].

The curve for graphite obtained by Thorn and Simpson [95] is indicated by 1. This is minimal for samples with ideally polished surfaces.

Samples with rough surfaces (curve 2) obtained by high-temperature vacuum sublimation were measured at 1480 and 2030°K; the spectral radiating power approached the limiting value of 0.91. Subsequently this maximum did not depend on temperature. The curve for carbon is indicated by 5. Jain et al. [96] (curves 3 and 4) used graphite samples, the surface of which was also subjected to "vacuum sublimation" at 2000°C. The samples used by other investigators were identical in origin and in type of surface treatment. As a rule, the surfaces were not carefully polished and the degree of roughness varied over wide limits.

In order to obtain values of total radiating power, various methods were used. Data on the integral radiating power for samples of noncrystalline carbon and graphite shown in Fig. 94 have quite a large scatter.

The results obtained by Jain et al. present practical interest, since the total and spectral radiating powers were measured simultaneously. The values of spectral radiating power show a negative variation with temperature (see Fig. 93, curves 3 and 4), while the total radiating power has a positive temperature coefficient. The curves approach one another at the maximum temperature.

As already mentioned above, the angular variation of the radiating power of a carbon filament obeys a cosine law of radiation distribution; thus the normal and hemispherical radiating powers are approximately equal. Since the optical properties of carefully polished graphite are similar to those of metals, we may expect these surfaces to show analogous deviations from Lambert's law.

Simultaneously with the study of graphite, many measurements were made of the reflecting power of lampblack, a material widely used to absorb radiation in optical systems and to blacken radiation receivers.

Royd [97] measured the reflecting power of soot and observed that the reflection coefficient varied from $\rho(\lambda) = 0.013$ at $\lambda = 0.8\ \mu$ and 0.0066 at $\lambda = 8.7\ \mu$ to 0.0067 at $\lambda = 25.5\ \mu$ and 0.016 at $\lambda = 51.0\ \mu$. Coblentz (98) made many measurements on the reflecting power of soot deposits from various kinds of fuel and observed that at $\lambda = 0.95\ \mu$ the reflection coefficient varied from 0.0064 to 0.0134 and at $\lambda = 24\ \mu$ from 0.030 to 0.057. The results of Coblentz' measurements are much higher than those of Royd and were made with greater accuracy. Harris and Cuff [99] measured the reflection and transmission coefficients of soot from $\lambda = 0.254$ to $\lambda = 1.100\ \mu$

and obtained results midway between those of Royd and Coblentz. Measurements of reflection from the deposit obtained on fuel combustion showed that the reflection coefficient rose with increasing wavelength. Pfund [100] observed that the spectral transmission of soot in the visible part of the spectrum was smaller than 0.1%, increasing on raising the temperature (0.15 for $\lambda = 3$ μ, 0.43 for $\lambda = 6$ μ, and 0.56 for $\lambda = 11$ μ). Plyler and Ball [101] obtained values of transmission for camphor smoke deposited on a gallium plate with bromoiodide windows: this gave 0.50 for $\lambda = 4$ μ, 0.65 for $\lambda = 8$ μ, 0.75 for $\lambda = 12$ μ, 0.75 to 0.80 for $\lambda = 3.2$ μ, and 0.70 for $\lambda = 4.0$ μ. These results also confirm that the transmission rises as wavelength increases.

Harris and Cuff [99] determined the scattering centers of the soot deposits as > 0.2 μ. Thus the scattering is the greater, the smaller the wavelength in comparison with the geometrical dimensions of the scattering center.

Sully et al. [102] obtained values of radiating power for soot covering metal heaters. Their results showed that the $\varepsilon_\Sigma(T)$ increases from 0.94 at 300°C to 0.99 at 500°C. For sufficiently thin layers the radiating power will be constant and quite high.

Measurements of Umur et al. [103] showed that the radiation of lampblack did not obey the cosine distribution law. For substantial angles from the normal there were great deviations. For an oxidized copper plate 3 mm thick covered with a layer of soot, the total normal radiating power was 0.95, and the hemispherical radiating power only 0.90.

Carbon and graphite arcs are normally used as sources of heat and radiation, and there is especially great interest in their radiating powers.

The interpretation of results for the $\varepsilon(\lambda, T)$ and $\varepsilon_\Sigma(T)$ of arcs is complicated by the fact that the radiating power of the plasma as well as that of the electrodes is measured.

Krijgsman [104] measured the spectral energy distribution of an arc and observed that the intensity of radiation corresponded to that of a black body.

MacPherson [105] came to the conclusion that an arc was a radiator forming a "gray body" with radiating power 0.91, but later this quantity was measured more accurately and proved to be 0.78.

Parker and Lock [106] measured the radial energy distribution of an arc in the range of 0.750 to 0.220 μ and came to the conclusion that the radiating power was close to 1.0.

Johnson [107] measured the ultraviolet spectral distribution in the range 0.250 to 0.190 μ and obtained data agreeing closely with the results of Parker and Lock for the longer wavelengths if the cross section of the laminated anode corresponded to the radiating power of a "gray body" equal to 0.96.

Null and Lozier [108] used an arc furnace for measuring the reflecting power of a carbon arc for wavelengths from 0.500 to 0.600 μ and obtained a radiating power of 0.98, in complete agreement with the values of Parker, Lock, and Johnson.

Warmuth [109] measured the reflecting power of a carbon arc at $\lambda = 0.663$ μ and found that $\varepsilon(\lambda) = 0.972$. This coincides with results obtained (for the same material) between room temperature and 1540°C.

Smaller values of radiating power were obtained at arc temperatures by Euler [89], who found that $\varepsilon(\lambda, T)$ = 0.70 to 0.75 in the wavelength range 0.400 to 0.700 μ. He concluded from this that the radiating power rose with increasing wavelength.

Anacker and Mannkopf [110] obtained radiating-power values of: $\varepsilon(\lambda, T) = 0.765$ at $\lambda = 0.482$ μ; $\varepsilon(\lambda, T) = 0.745$ at $\lambda = 400$ μ; and $\varepsilon(\lambda, T) = 0.775$ at $\lambda = 0.640$ μ.

The different values of the radiating powers of carbon and graphite at arc temperatures may be explained by their dependence on the porosity and the degree of surface finish due to the powder covering the surface of the electrode. Apparently the radiation of the plasma also tends to increase the value of the radiating power. The results of Johnson [107] indicate this effect. In individual measurements, the plasma radiation also affects the value of radiating power obtained in the ultraviolet region.

Thus, as we may conclude from the foregoing data, the radiating power of carbon and graphite are determined by a combination of properties.

Careful study of the various experimental results enabled Plankett and Kingery to conclude (similar conclusions were also drawn by MacPherson) that:

1. The spectral and total radiating powers of carbon and graphite in the temperature range 1000 to 2000°C are fairly high in comparison with those of metals;
2. The spectral radiating power falls with increasing wavelength for $\lambda > 1\,\mu$, and then increases; in the infrared part of the spectrum graphite and carbon cannot be regarded as "graphite bodies";
3. The spectral radiating power of carbon and graphite in the visible region is the same as that of a "gray body";
4. The maximum value of reflecting power at room temperature for polished graphite lies at $\lambda = 0.260\,\mu$;
5. The spectral radiating power at $\lambda = 0.665\,\mu$ for carbon and graphite varies from ~ 0.77 for a polished surface to 0.95 for coarse, rough surfaces; the temperature coefficient remains the same;
6. The values of spectral radiating power for carbon and graphite may vary from 0.70 to 0.90;
7. The "normal" radiating power exceeds the "hemispherical" by some 5% for coarse surfaces (this value rises for well-polished surfaces);
8. The transmission of soot-covered plates increases markedly for long waves $\lambda > 5\,\mu$;
9. The radiating power in arc processes is only 2 to 3% smaller than the radiation of a black body in the visible and ultraviolet regions.

The most careful measurements of spectral and total radiating power of graphite were made by Plankett and Kingery [88]. The results of this work form the basis of the present chapter.

The authors paid special attention to preparing and treating the sample surfaces. The samples were prepared from graphite and carbon manufactured by the Nation Carbon Company (USA). Plane samples 3.6 mm thick and some 300 mm long were cut into sections with transverse diameter ~ 100 mm. Then they were worked so as to acquire a length of 72 mm and thickness approximately 24 mm. One end was carefully cut off. After final treatment (when the samples took the form of disks), the axis of the cylinder was set at 90° to its plane. This made it possible to orient the maximum number of graphite crystals so that their axes should also lie at 90° to the surface of the cylindrical sample. In the samples made in the form of disks, the surface finish involved different grain sizes. The polished samples were made by the Coblentz method: The sample was wetted with water and ground by special disks, then placed in a furnace until a smooth, lustrous surface resulted.

A sample with a very coarse surface was obtained by placing for 12 min in an air furnace at 900°C. After the sample was taken from the furnace, it was treated in a special sand-blasting apparatus in order to remove oxide scale formed during heating.

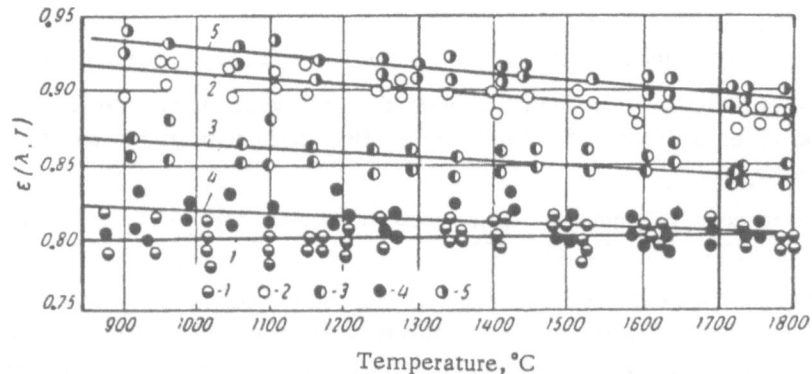

Fig. 95. Variation of the $\varepsilon(\lambda, T)$ of graphite ($\lambda = 0.665\,\mu$) with temperature. 1-4) Polished samples; 5) sample oxidized and sand-blasted.

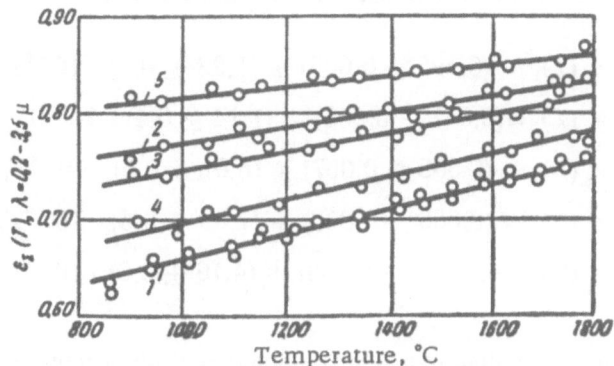

Fig. 96. Total radiating power of graphite for the spectral range 0.2 to 3.5 μ (notation of samples as in Fig. 95).

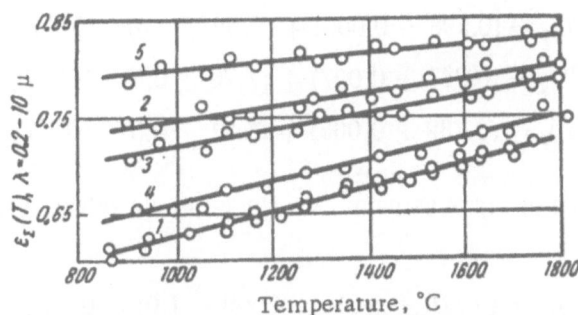

Fig. 97. Total radiating power of graphite for the spectral range 0.2 to 10.0 μ (notation as in Figs. 95 and 96).

Pyrolytic graphite samples were prepared by depositing pyrolytic graphite on graphite disks at 2100°C. The thickness of sample 1P, prepared by the coating method, was 200 to 300 μ, that of sample 2P was 30 to 70 μ, and 3P 40 to 60 μ; sample 4P was polished to 4 to 6 μ.

Figure 95 shows results obtained for the spectral radiating power of graphite samples (National Carbon Company).

Since the relationships formed straight lines, the following expressions were obtained by the method of least squares* for λ = 0.665 μ:

$$\varepsilon(\lambda, T)_1 = (0.803 \pm 0.06) - (2.49 \pm 0.5) \cdot 10^{-6} T,$$
$$\varepsilon(\lambda, T)_4 = (0.849 \pm 0.009) - (2.38 \pm 0.7) \cdot 10^{-5} T,$$
$$\varepsilon(\lambda, T)_3 = (0.896 \pm 0.008) - (2.71 \pm 0.6) \cdot 10^{-5} T,$$
$$\varepsilon(\lambda, T)_2 = (0.962 \pm 0.009) - (3.96 \pm 0.7) \cdot 10^{-5} T,$$
$$\varepsilon(\lambda, T)_5 = (0.975 \pm 0.001) - (4.03 \pm 0.7) \cdot 10^{-5} T,$$

where T is the Kelvin temperature (we should note that the graphs were taken with intervals of 100°K).

*Samples 2, 3, and 4 had respective grain sizes 80, 180, and 320; sample 1 was polished, sample 5 oxidized.

Figure 96 shows graphs for the total radiating power of the same samples in the range 0.2 to 3.5 μ:

$$\varepsilon_{\Sigma}(T)_1 = (0.495 \pm 0.005) + (1.22 \pm 0.3) \cdot 10^{-4}T,$$

$$\varepsilon_{\Sigma}(T)_4 = (0.577 \pm 0.006) + (1.14 \pm 0.4) \cdot 10^{-5}T,$$

$$\varepsilon_{\Sigma}(T)_3 = (0.663 \pm 0.007) + (8.81 \pm 0.5) \cdot 10^{-4}T,$$

$$\varepsilon_{\Sigma}(T)_2 = (0.668 \pm 0.006) + (7.89 \pm 0.5) \cdot 10^{-5}T,$$

$$\varepsilon_{\Sigma}(T)_5 = (0.768 \pm 0.006) + (4.16 \pm 0.6) \cdot 10^{-5}T.$$

Figure 97 shows that the total radiating power as a function of temperature in the wavelength range 0.2 to 10.0 μ. The corresponding equations are

$$\varepsilon_{\Sigma}(T)_1 = (0.448 \pm 0.07) + (1.38 \pm 0.4) \cdot 10^{-4}T,$$

$$\varepsilon_{\Sigma}(T)_4 = (0.504 \pm 0.08) + (1.19 \pm 0.6) \cdot 10^{-4}T,$$

$$\varepsilon_{\Sigma}(T)_3 = (0.598 \pm 0.006) + (9.62 \pm 0.5) \cdot 10^{-4}T,$$

$$\varepsilon_{\Sigma}(T)_2 = (0.647 \pm 0.007) + (7.76 \pm 0.5) \cdot 10^{-5}T,$$

$$\varepsilon_{\Sigma}(T)_5 = (0.732 \pm 0.005) + (5.21 \pm 0.5) \cdot 10^{-5}T.$$

Analogous measurements for pyrolytic graphite gave the following values of spectral radiating power for $\lambda = 0.665\ \mu$ for the four samples:

$$\varepsilon(\lambda, T)_{1P} = (0.776 \pm 0.016) - (2.2 \pm 1.0) \cdot 10^{-6}T,$$

$$\varepsilon(\lambda, T)_{2P} = (0.812 \pm 0.017) - (0.91 \pm 1.2) \cdot 10^{-6}T,$$

$$\varepsilon(\lambda, T)_{3P} = (0.833 \pm 0.015) - (1.02 \pm 0.9) \cdot 10^{-6}T,$$

$$\varepsilon(\lambda, T)_{4P} = (0.917 \pm 0.015) - (1.11 \pm 0.9) \cdot 10^{-6}T$$

and correspondingly for the total radiating power in the range 0.2 to 3.5 μ:

$$\varepsilon_{\Sigma}(T)_{1P} = (0.512 \pm 0.02) + (1.17 \pm 1.0) \cdot 10^{-4}T,$$

$$\varepsilon_{\Sigma}(T)_{2P} = (0.526 \pm 0.018) + (1.4 \pm 0.9) \cdot 10^{-4}T,$$

$$\varepsilon_{\Sigma}(T)_{3P} = (0.587 \pm 0.016) + (1.35 \pm 0.7) \cdot 10^{-4}T,$$

$$\varepsilon_{\Sigma}(T)_{4P} = (0.641 \pm 0.015) + (1.17 \pm 0.7) \cdot 10^{-4}T.$$

For the total radiating power in the range 0.2 to 10 μ the same samples gave the values

$$\varepsilon_{\Sigma}(T)_{1P} = (0.473 \pm 0.013) + (1.27 + 0.6) \cdot 10^{-4}T,$$

$$\varepsilon_{\Sigma}(T)_{2P} = (0.506 \pm 0.015) + (1.36 \pm 0.7) \cdot 10^{-4}T,$$

$$\varepsilon_{\Sigma}(T)_{3P} = (0.561 \pm 0.012) + (1.46 \pm 0.4) \cdot 10^{-4}T,$$

$$\varepsilon_{\Sigma}(T)_{4P} = 0.634 \pm 0.016) + (1.12 \pm 0.8) \cdot 10^{-4}T.$$

Measurements for pure carbon (National Carbon Company) appear in Fig. 98. Here we have the variation of spectral radiating power for surfaces of two types. The equation for the smooth surface (white circles) is as follows:

$$\varepsilon(\lambda, T) = (0.862 \pm 0.008) - (3.57 \pm 0.6) \cdot 10^{-5}T$$

and that for the rough surface (black circles):

$$\varepsilon(\lambda, T) = (0.0892 \pm 0.009) - (4.21 \pm 0.6) \cdot 10^{-5}T.$$

Expressions were obtained for the total radiating powers of the same two samples in the range 0.2 to 3.5 μ.

The results appear in Fig. 99:

For the smooth surface

$$\varepsilon_{\Sigma}(T) = (0.789 \pm 0.007) - (5.5 \pm 0.6) \cdot 10^{-6}T$$

For the rough (coarse) surface

$$\varepsilon_{\Sigma}(T) = (0.822 \pm 0.008) - (1.79 \pm 0.6) \cdot 10^{-6}T.$$

The expressions for the total radiating power in the range 0.2 to 10 μ appear in Fig. 100:

For the smooth surface

$$\varepsilon_{\Sigma}(T) = (0.781 \pm 0.009) - (1.30 \pm 0.7) \cdot 10^{-6}T$$

For the coarse surface

$$\varepsilon_{\Sigma}(T) = (0.823 \pm 0.008) - (1.29 \pm 0.6) \cdot 10^{-6}T.$$

Tables 22 to 24 give values of the resultant radiating and reflecting powers measured at room temperature for various surface finishes. The reflecting power was measured in the spectral range 0.400 to 0.700 μ.

Plankett and Kingery's results enables them to reach some interesting conclusions stemming directly from the above expressions, graphs, and tables, and reducing in brief to the following. The increase in the radiating power of graphite with increasing grain size (defined by the dimensions of the "crystallites") is not always observed. Thus the radiating power in sample 4 was smaller than in sample 3, and in sample 3 smaller than in sample 2. Plankett and Kingery seek the explanation for this in the orientation of the crystals, drawing an analogy between the crystal structure of graphite and that of a metal.

From this point of view, for the range of wavelengths used, the radiation surface of the sample with large crystal faces (for example, 4) is "less rough" than those with small ones (sample 3 and 2, respectively). The smallest value of radiating power for the sample with the polished surface and the largest for the sample with the oxidized surface are evident.

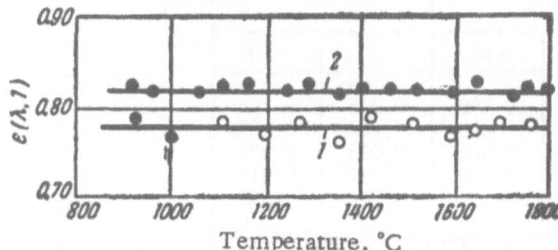

Fig. 98. Spectral radiating power of graphite (λ = 0.665 μ) for samples. 1) With smooth surface; 2) with rough surface.

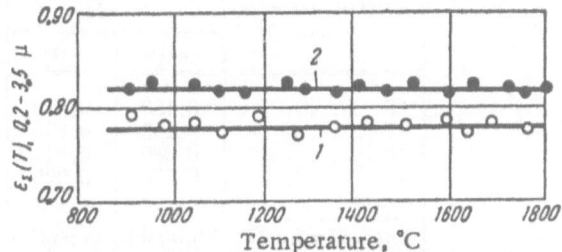

Fig. 99. Total radiating power of carbon in the spectral range 0.2 to 3.5 μ for samples. 1) With smooth surface; 2) with rough surface.

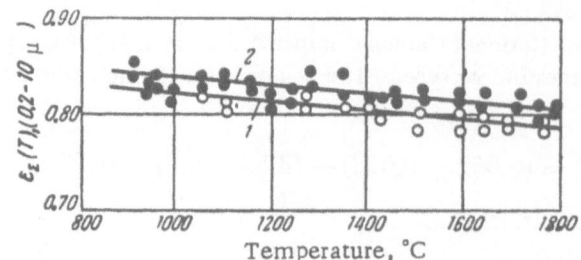

Fig. 100. Total radiating power for carbon in the spectral range 0.2 to 10 μ for samples. 1) With smooth surface; 2) with rough surface.

The temperature coefficient of the spectral radiating power of graphite is negative, i.e., coincides in sign with the temperature coefficient of the specific electrical resistance of graphite. This enables us to explain the fall in the spectral radiating power with temperature qualitatively by Drude's formula. In the above-cited work of Thorn and Simpson, however, the temperature coefficient is positive, which is impossible to explain by Drude's formula, even qualitatively. But this is not surprising, and was already noted in considering the analogous question for metals.

TABLE 22

Graphite sample	For wavelength λ, μ					
	0.400		0.665		0.700	
	$\rho(\lambda)$	$\varepsilon(\lambda)$	$\rho(\lambda)$	$\varepsilon(\lambda)$	$\rho(\lambda)$	$\varepsilon(\lambda)$
1, polished	0.175	0.825	0.190	0.810	0.195	0.805
4, grain size 320	0.170	0.830	0.178	0.822	0.180	0.820
3, grain size 180	0.122	0.878	0.125	0.875	0.128	0.872
2, grain size 80	0.102	0.898	0.104	0.896	0.105	0.895
5, oxidized	0.063	0.937	0.064	0.036	0.066	0.934

TABLE 23

Carbon sample	For wavelength λ, μ					
	0.400		0.665		0.700	
	$\rho(\lambda)$	$\varepsilon(\lambda)$	$\rho(\lambda)$	$\varepsilon(\lambda)$	$\rho(\lambda)$	$\varepsilon(\lambda)$
Rough	0.090	0.910	0.090	0.910	0.090	0.910
Smooth	0.065	0.935	0.067	0.933	0.070	0.930

TABLE 24

Pyrolytic-graphite sample	For wavelength λ, μ					
	0.400		0.665		0.700	
	$\rho(\lambda)$	$\varepsilon(\lambda)$	$\rho(\lambda)$	$\varepsilon(\lambda)$	$\rho(\lambda)$	$\varepsilon(\lambda)$
1P	0.190	0.810	0.210	0.790	0.215	0.785
2P	0.145	0.855	0.160	0.840	0.165	0.835
3P	0.089	0.911	0.092	0.908	0.093	0.907
4P	0.075	0.925	0.080	0.920	0.082	0.918

Note: Sample 1P obtained by deposition (200 to 300 μ); sample 2P mechanically treated (30 to 70 μ); sample 3P mechanically treated (40 to 60 μ); sample 4P polished metallographically (4 μ).

TABLE 25

Samples	For wavelength λ, μ		
	extrapolated values of $\varepsilon(\lambda, T)$	measured $\varepsilon(\lambda, T)$	difference $\Delta\varepsilon(\lambda, T)$
Carbon	0.852	0.910	—0.058
»	0.879	0.932	—0.053
Graphite			
1, polished · · · · · · · · · · · ·	0.802	0.810	—0.018
4, grain size 320 · · · · · · · · · ·	0.842	0.822	+0.020
3, grain size 180 · · · · · · · · · ·	0.888	0.875	+0.013
2, grain size 80 · · · · · · · · · ·	0.950	0.897	+0.053
5, oxidized . · · · · · · · · · ·	0.963	0.936	+0.027
Pyrolytic graphite			
1P	0.775	0.790	—0.015
2P	0.812	0.840	—0.028
3P	0.882	0.910	—0.028
4P	0.914	0.920	—0.006

The differences in temperature dependence may be explained by the anisotropy of the optical properties of graphite, which may be expected in view of the anisotropy of its electrical properties.

The dependence of the total radiating power on temperature is positive, and this is turn inclines one to the view that the temperature coefficient of the spectral radiating power is close to zero. For carbon this view is undoubtedly valid.

The negative slope of the straight lines giving the total radiating power in carbon in the wavelength range 0.2 to 10 μ is so slight that it practically lies within the limits of experimental error.

Data on the extrapolation of spectral reflecting powers (measured by reflection at room temperature) to high temperatures, also carried out by Plankett and Kingery, are shown in Table 25. These results indicate an extremely small value of the temperature coefficient for the spectral radiating power of graphite. However, the high values of the temperature coefficient at arc temperatures make us treat this kind of extrapolation with caution.

In conclusion, we present comparative data on the spectral reflecting power of graphite and diamond obtained by Ercun et al. [111] for the visible part of the spectrum at room temperature. These data are shown in Fig. 101. Here curves 1 and 2 correspond to the maximum or minimum values of $\rho(\lambda, T)$ for graphite. Curve 3 corresponds to the spectral distribution (λ) for diamond. The similarity between the minimum values of $\rho(\lambda)$ for graphite (curve 2) and the diamond values is very striking

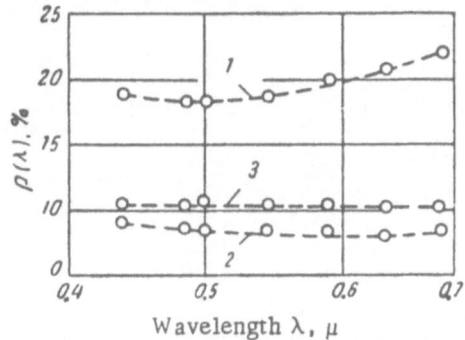

Fig. 101. Spectral reflecting power of graphite.
1) Maximum values; 2) minimum values; 3) for diamond (measured by the oil-immersion method).

CHAPTER VI

EXPERIMENTAL DATA ON THE RADIATING POWER
OF CERTAIN OXIDES AND CARBIDES

Certain Ceramic Compounds

Results of measuring $\rho(\lambda, T) = 1 - \varepsilon(\lambda, T)$ for refractories composed of silica, aluminum oxide, and mixtures of these obtained by Allegre [112] at 1000°C in the modulation reflectometer are shown in Figs. 102-104. The composition of the sample of Fig. 102 was 95% SiO_2, 3% CaO, and 2% kaolin. That of the sample of Fig. 103 was 55.7% SiO_2, 41.4% Al_2O_3, 2% Fe_2O_3, and 0.5% CaO.

Figure 104 shows the reflecting power $\rho(\lambda, T)$ of a sample consisting of more than 95% Al_2O_3. One notices the "gray" character of the radiation from these refractories in the near-infrared part of the spectrum in the region of 5 to 7 μ for the first, 6 to 8 for the second, and 8 to 9 μ for the third sample. These results are in good agreement with those of the same author given in Chapter II.

The total radiating power of pure oxides in the form of specially prepared layers was studied by Sully et al. [102], the oxide films being deposited on a metal-strip substrate serving as heater. Figure 105 shows the relationship for $\varepsilon_\Sigma(T)$ in the case of 1) silicon, 2) thorium, 3) magnesium, 4) zirconium, 5) cerium, and 6) aluminum oxides. Data on the $\varepsilon_\Sigma(T)$ of aluminum oxide are presented as a function of film thickness, measured in grams of oxide per 1 cm^2 surface, in Fig. 106.

As we see, for large coating thicknesses, when the film "approaches" a massive sample, the results of this work correspond all the more to those of Allegre quoted above.

For thin coatings (Fig. 106) and high temperatures, the results are distorted by the radiation of the metal substrate, which penetrates through the film under examination. For small thicknesses, this film partly transmits radiation.

All this is confirmed by Fig. 107, which presents data of the same authors for pure cerium oxide deposited on a substrate of Nimonic alloy. Various oxide-coating thicknesses were deposited.

Here curve 1 corresponds to pure cerium oxide deposited on a substrate composed of a previously oxidized Nimonic-alloy surface; curve 2 corresponds to the same surface polished. For a coating thickness of $\sim 0.03 \mu$, the radiation from the substrate fails to penetrate the oxide layer [results on the $\varepsilon_\Sigma(T)$ of the polished and oxidized surfaces identical].

Results for the $\varepsilon_\Sigma(T)$ and $\varepsilon(\lambda, T)$ of the oxides of rare earths for wavelengths $\lambda = 0.640 \mu$ and $\lambda = 1 \mu$ are given below from data of Blair [113]. It should be noted, however, that the data obtained by this author for

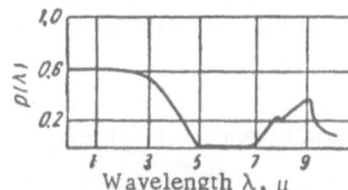

Fig. 102. Spectral reflection coefficient of refractories (95% SiO_2, 3% CaO, and 2% kaolin).

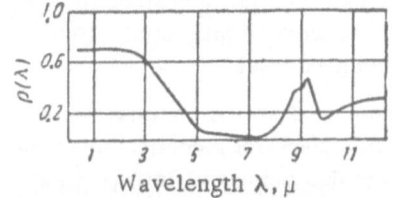

Fig. 103. Spectral reflection coefficient of refractories (55.7% SiO_2, 41.4% Al_2O_3, 2% FeO, and 0.5% CaO.

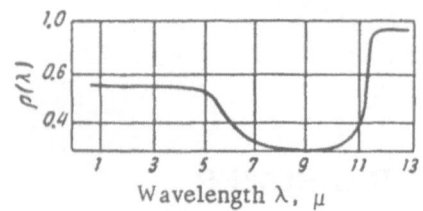

Fig. 104. Spectral reflection coefficient of a ceramic surface ($Al_2O_3 > 95\%$).

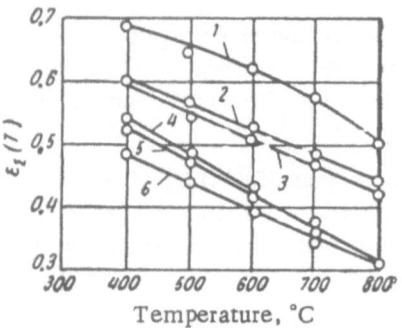

Fig. 105. Variation of the $\varepsilon_\Sigma(T)$ of pure oxides.

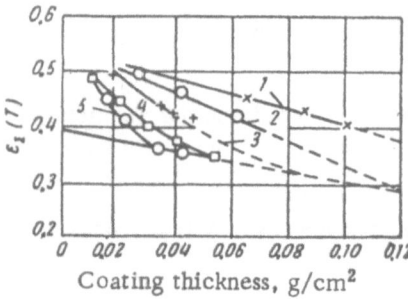

Fig. 106. Variation of the $\varepsilon_\Sigma(T)$ of aluminum oxide with coating thickness. Oxide-particle dimensions, respectively. 1) $\lambda = 44\ \mu$; 2) $\lambda = 30\ \mu$; 3) $\lambda = 20\ \mu$; 4) $\lambda = 13\ \mu$; 5) $\lambda = 8\ \mu$.

the spectral radiating power (at $\lambda = 1.0\ \mu$) and $\varepsilon_\Sigma(T)$ of industrial refractories composed of aluminum oxide raise some doubts.*

Figure 108 gives data for gadolinium oxide. Curve 1 corresponds to the $\varepsilon_\Sigma(T)$ in the range $\lambda = 0.3$ to $2.7\ \mu$, curve 2 gives $\varepsilon(\lambda, T)$ for $\lambda = 0.640\ \mu$ and curve 3 gives $\varepsilon(\lambda, T)$ for $1.0\ \mu$. Analogous curves are given for erbium oxide in Fig. 109 and samarium oxide in Fig. 110.

These data (as far as we know the only ones for rare-earth oxides) are not absolutely clear as regards the $\varepsilon(\lambda, T)$ for $1\ \mu$ and to a certain extent the $\varepsilon_\Sigma(T)$; they require confirmation.

In considering the radiating power of oxides, we must mention the extremely important phenomenon of nonthermal luminescence which may be observed especially on flame heating. Thus in Sokolov et al. [114], pp. 773-774, the existence of "selective radiation" for calcium and magnesium oxides placed in a luminous-gas flame is demonstrated. The reason for this phenomenon of chemiluminescence is to be found in the exothermic processes of the recombination of free atoms and radicals on the sample surface, with the formation of molecules.

At temperatures exceeding the luminescence-quenching temperature of the sample, and at any other temperatures in the case of luminophores, the energy of recombination passes mainly into thermal energy of the sample. It was established that the additional heating in the luminous-gas flame was 75°C for calcium oxide and 30°C for magnesium oxide. This corresponds to recombination coefficients of 0.4 to 1.0.

Calculation shows that at 1000°K the additional heating increases the radiation in the visible part of the spectrum three times for magnesium oxide and nine times (!) for calcium oxide.

The spectral and total radiating powers of the oxidized surfaces of metals and alloys have been studied by many workers.

In view of the excessively large number of variables, including the surface finish, the thermal conditions of forming various oxide films, and so on, we feel that there is not much point in comparing the data of different papers.

It must also be emphasized that the radiating powers of oxide films to a large extent depend on the actual process (time of oxidation, etc.). In view of the importance of this question for a number of processes, the final section of this chapter is devoted to information on $\varepsilon(\lambda, T)$ and $\varepsilon_\Sigma(T)$ in the "dynamics of development" (appearance or disappearance) of the film. Here we present some results of practical interest, which are also fairly characteristic of "stably-oxidized" coatings. Thus Fig. 111 gives data from the above-mentioned work of Sully et al. [102] on the $\varepsilon_\Sigma(T)$ for Nimonic 75 alloy, Fig. 112 is for nickel, and Fig. 113 for stainless steel.

The samples of all the materials were carefully cleaned, then they were oxidized in electric furnaces with an air atmosphere for a considerable time. Naturally oxidation substantially reduces the differences in radiating power due to surface finish (see Fig. 111).

*Thus for $\lambda = 1\ \mu$ at 1000°C Blair obtained $\varepsilon(\lambda, T) = 0.85$ to 0.95, and for $\varepsilon_\Sigma(T)$ the range 0.3 to 2.7 μ he obtained not a fall but a rise (?) in the total radiating power on increasing the temperature.

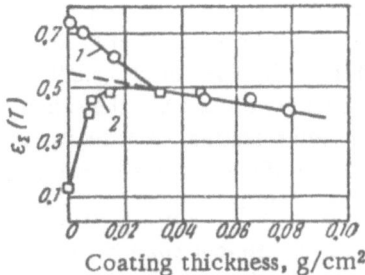

Fig. 107. Variation of the $\varepsilon_\Sigma(T)$ of cerium oxide with thickness of coating deposited on an oxidized metal substrate

This is clearly seen for all materials. Thus Fig. 111a corresponds to a rough surface after sand-blasting, Fig. 111b to a polished, and Fig. 111c to a rolled surface.

In all these figures the figure 1 represents $\varepsilon_\Sigma(T)$ curves for oxidation at 1200°C, 2 for 900°C, and 3 for 600°C; figure 4 indicates curves for an unoxidized surface. Figure 112a and b correspond to unpolished (after sand-blasting) and polished nickel surfaces. Curves 1 correspond to oxidation at 900°C, 2 to 600°C, while 3 corresponds to the unoxidized nickel surface. Somewhat unexpected is the sharp rise in $\varepsilon_\Sigma(T)$ for oxidation of unpolished nickel at 600°C. In stainless-steel samples (Fig. 113) sand-blasted (Fig. 113a), polished (Fig. 113b), and rolled (Fig. 113c), the figures 1, 2, 3, as for nickel, respectively, indicate oxidation at 900 and 600°C and a pure unoxidized surface.

Values of $\varepsilon_\Sigma(T)$ as a function of emission angle φ for oxidized Nimonic and nickel surfaces are given in Table 26. For comparison we also show values for a pure aluminum mirror.

As we see, the cosine law is practically obeyed for oxidized Nimonic and nickel surfaces.

Carbides

The radiating power of carbides, the most high-melting of materials, is only now being studied. The fullest data on the carbides of zirconium and tantalum were obtained by Riethof et al. [49].

The normal spectral relationship of tantalum carbide (Fig. 114) was measured in an argon atmosphere. The point with zero temperature coefficient, x, lies in the red part of the spectrum ($\lambda \approx 0.65 \, \mu$) for this carbide.

It should be noted that these data on the $\varepsilon(\lambda, T)$ of tantalum carbide sharply contradict those published earlier by Blau [115].

The disagreement between these values may well be due to the presence of tantalum pentoxide on the sample surfaces [114]. The presence of this in Blau's work was indicated by x-ray analysis. Still unexplained are not only the reasons for the "overestimated" absolute values of $\varepsilon(\lambda, T)$ for tantalum carbide in the infrared part of the spectrum, which may be connected with the type of surface finish, but also the "sharp anomaly" in the behavior of the spectral radiating power on increasing the temperature and wavelength. Bearing in mind what we have said, the reader may acquaint himself with the results of this work from Fig. 115.

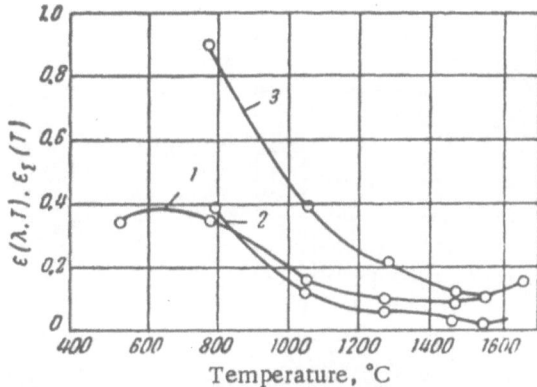

Fig. 108. Temperature dependence of the radiating power of gadolinium oxide: 1) Total; 2) spectral for $\lambda = 0.640 \, \mu$; 3) spectral for $\lambda = 1.0 \, \mu$.

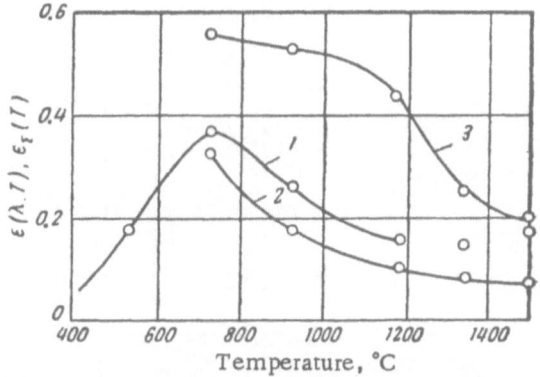

Fig. 109. Temperature dependence of the total and spectral radiating power of erbium oxide (notation as in Fig. 108).

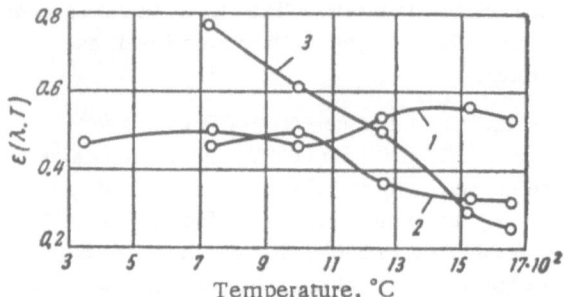

Fig. 110. Temperature dependence of the total and spectral radiating power of samarium oxide (notation as in Figs. 108 and 109).

Data on the spectral normal radiating power of zirconium carbide, also obtained in the work of Riethof et al. [49], are shown in Fig. 116. The samples studied contained up to 96% ZrC. Before measuring, the samples were aged, both before and after heating, so as to attain a "stable" surface state. The surface was monitored by microphotography and x-ray analysis.

The zero-temperature-coefficient point X for zirconium carbide lies at wavelength $\sim 2.4\ \mu$.

From the measured data on the normal spectral radiating power, Riethof et al. [49] calculated and plotted the normal total radiating power ε_Σ (T). The results of these calculations appear in Fig. 117 in the form of ε_Σ(T) curves for zirconium and tantalum carbides. The later curve coincided with the ε_Σ(T) curve for tungsten calculated from the $\varepsilon(\lambda$, T) data in Fig. 58 for metallic molybdenum. Data on the normal spectral radiating power of tungsten and molybdenum are given above.

TABLE 26

Material	φ°	D	K
Oxidized Nimonic	0 20 40 60 80	13.35 12.70 10.95 7.84 3.69	39.3 38.3 38.6 38.8 38.8
Oxidized nickel	0 20 40 60 80	11.00 10.70 9.10 6.25 2.65	33.4 32.2 33.0 31.9 28.8
Oxidized aluminum	0 60	8.15 5.05	24.0 25.0

Note. D is a quantity proportional to the intensity of radiation. The constancy of $K = \dfrac{D}{a \cos \varphi + b \sin \varphi}$ (where a and b are constants) indicates obedience to Lambert's cosine law.

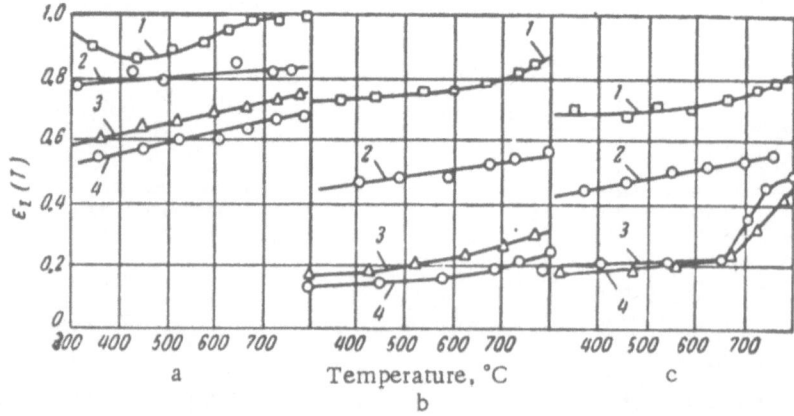

Fig. 111. Temperature dependence of ε_Σ(T) for a Nimonic 75 surface.

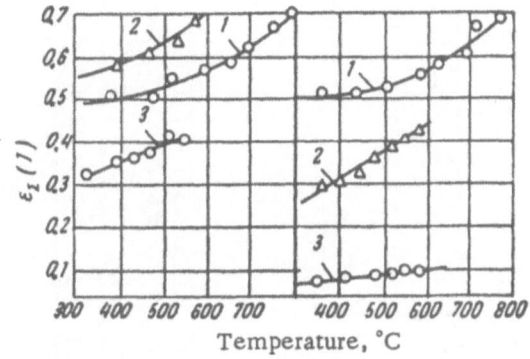

Fig. 112. Temperature dependence of ε_{Σ} (T) for a nickel surface.

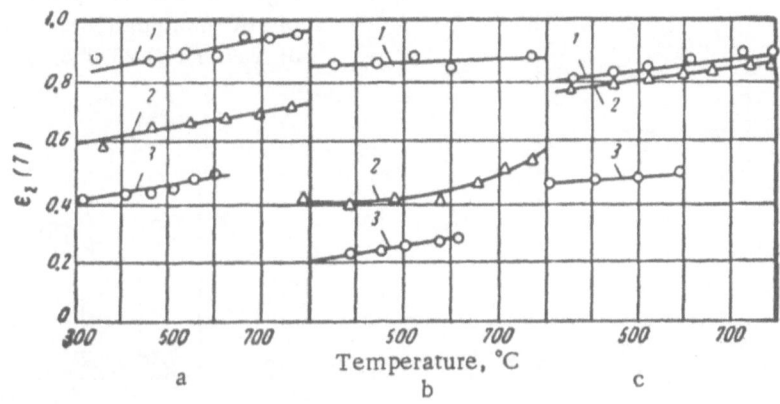

Fig. 113. Temperature dependence of ε_{Σ} (T) for a stainless-steel surface.

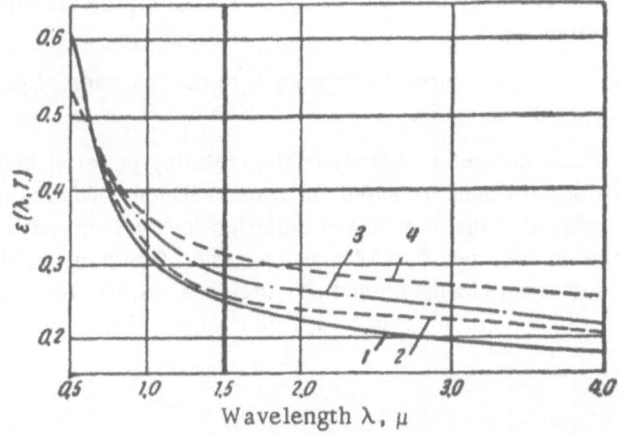

Fig. 114. Normal spectral radiating power of tantalum carbide. 1) 1830°K; 2) 2250°K; 3) 2670°K; 4) 2880°K.

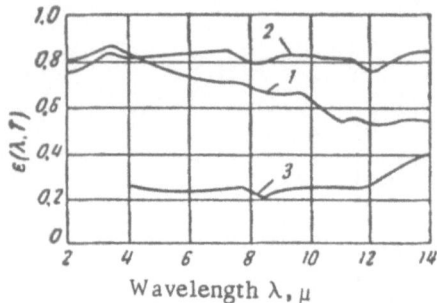

Fig. 115. $\varepsilon(\lambda, T)$ of tantalum carbide.
1) 1950°K; 2) 2250°K; 3) 3234°K.

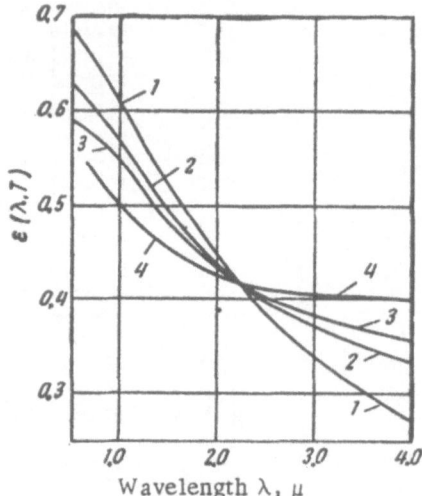

Fig. 116. Normal spectral radiating power
of zirconium carbide. 1) 2100°K; 2) 2270°K;
3) 2470°K; 4) 2670°K.

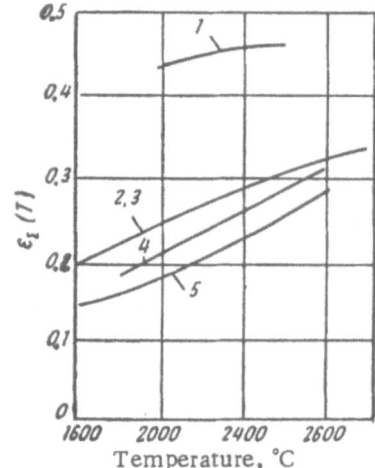

Fig. 117. Total normal radiating power of
1) zirconium carbide; 2) tantalum carbide;
3) tungsten; 4) tantalum; 5) molybdenum.

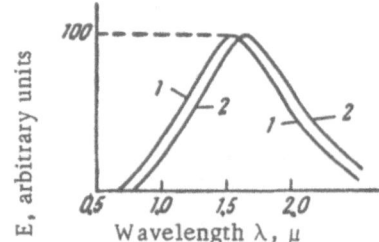

Fig. 118. Radiation of tantalum
and tungsten. 1) Pure; 2) carbidized (6% C).

The effect of the "carbidization" process on the radiating power of tungsten [116] reduces to a displacement of the curve giving the spectral energy distribution of the radiation, the maximum of this curve shifting
into the infrared part of the spectrum.

This is illustrated by Fig. 118, where curve 1 corresponds to the radiation of pure tantalum or tungsten and
curve 2 to the carbidized variety (containing 6% C).

Many investigations have been devoted to a study of the radiating power of high-melting compounds of
the "globar" type, the compositions of which are based on silicon carbide. A fairly full bibliography is given
in Samsonov [117]. It is here shown that the intensity of radiation for the compositions silicon carbide, silicon
nitride, and molybdenum silicide in the range 4 to 14 μ is the same. The position of the maximum in the emission spectrum of silicon carbide does not change on adding molybdenum silicide [3 to 14%(mol.)]. Addition
of silicon nitride [6%(mol.)] to silicon carbide leads to a displacement of the radiation maximum in the short-
wave direction.

Glass

We considered the main questions connected with the transmission and radiation of glass earlier.

Figures 119 and 121 give data of Sterlyadkina [118] on the spectral absorption of glass as a function of iron-oxide content in the near-infrared part of the spectrum for temperatures 600 to 1300°C. This work was done by measuring the flux passing through a layer of molten glass and reflected from a platinum mirror placed beneath it.

The ordinates in Figs. 119-121 represent the absorption K in cm^{-1}:

$$K = \frac{\lg \frac{I_1}{I_2}}{\lg e \, (d_2 - d_1)},$$

where I_1 and I_2 are the radiant fluxes and $(d_2 - d_1)$ is the thickness of the glass layer.

Radiating Power of Some Metallic Surfaces in the Dynamics of the Oxidation

Process

We mentioned above that the radiating power of oxide films was determined to a considerable extent by the dynamics of the actual oxidation process. In other words, in the course of one kind of process or another, the radiating power changes not only as a result of changing temperature but also because of reactions taking place on the surface itself.

As typical examples of such processes we shall present data obtained by Svet [119] on the $\varepsilon(\lambda, T)$ of a steel bath in the near-infrared part of the spectrum. The study was made by means of a partial-radiation pyrometer with effective wavelength $\lambda \approx 1.5 \, \mu$, while the true temperatures of the steel bath were measured by an immersion thermocouple.

The radiating power of the steel bath was also studied on adding up to 0.6% V and 1.3% Cr.

Examination of the changes in the radiating power of the steel bath for small additions of vanadium and chromium showed that the effect of the alloying elements on the value of $\varepsilon(\lambda, T)$ reduced in parctice to a change in the surface of the bath (appearance of a film).

The effect of adding vanadium on the radiating power may be judged from Fig. 122. We see on the graph giving the variance of $\varepsilon(\lambda, T)$ that in the high-temperature range (region A) the radiating power of the bath containing vanadium does not differ from the earlier-determined value without vanadium, since in this region the temperature is quite high, and the vanadium-dioxide film forming on the surface of the metal is dissolved by the slag and is taken up by the walls of the crucible (as a result of the mixing taking place in the induction furnace), i.e., the pyrometer signts the pure metal surface. At temperatures below 1600°C (region B), the bath begins to be covered with a thin film of undissolved oxide, and the radiating power grows very rapidly.

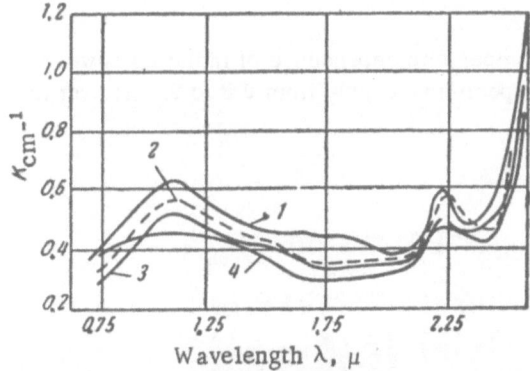

Fig. 119. Spectral absorption of glass containing 0.15% FeO at various temperatures. 1) 600°C; 2) 800°C; 3) 1000°C; 4) 1300°C.

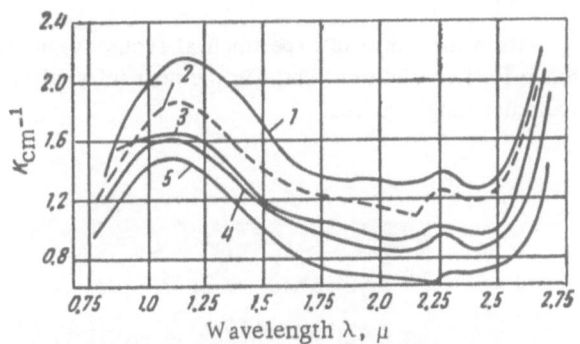

Fig. 120. Spectral absorption of glass containing 0.27% FeO at various temperatures. 1) 600°C; 2) 800°C; 3) 1300°C; 4) 1000°C; 5) 1200°C.

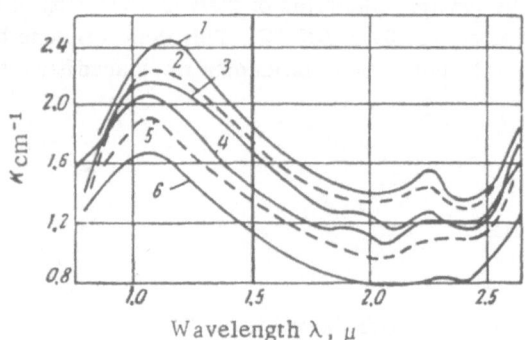

Fig. 121. Spectral transmission of glass containing 0.47% FeO at various temperatures. 1) 600°C; 2) 800°C; 3) 1350°C; 4) 1000°C; 5) 1300°C; 6) 1200°C

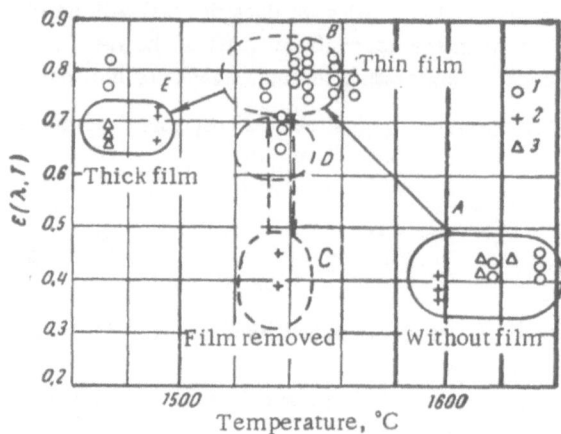

Fig. 122. $\varepsilon(\lambda, T)$ of a steel bath.

The radiating power in fact reaches the order of 0.8, i.e., rises by 100%. Here we must remember that, in these experiments, the changes in the radiating power of the bath on adding alloying elements were studied for a number of melts. Despite slight differences in chemical composition, the results obtained were identical. The points plotted on the graph (Fig. 122) correspond to the following melts: 1) 0.33% C, 0.03% Si, 0.46% Mn, 0.019% S, 0.56% V; 2) 0.32% C, 0.05% Si, 0.44% Mn, 0.16% S, 0.52% V; 3) 0.3% C, 0.16% Si, 0.49% Mn, 0.53% V (in all three melts the initial vanadium content was ~ 0.6%, and the metal also contained ~ 0.1% Ni.)

In order to check the effect of the film on the radiating power, the film was removed mechanically. The radiating power thereupon slowly fell by a factor of two. This simple but very convincing experiment is shown in Fig. 122 as the unstable region C. The arrow BC indicates the fall in $\varepsilon(\lambda, T)$ for "forced" removal of the film, and the arrow CD the increase in $\varepsilon(\lambda, T)$ as a result of the appearance of a new film on the bath surface.

Some fall in the radiating power on further film growth (region E) and on keeping the bath isothermal (region D) is only apparent. In fact, owing to the growth of the film, which constitutes a thin but perceptible skin of vanadium spinel with low thermal conductivity, a temperature drop develops between the bath and the skin surface. This drop leads to error in measuring the temperature, since the thermocouple measures the temperature of the molten bath, the surface of which is protected by the poorly conducting skin, while the pyrometric tube "sees" the surface temperature of this skin. Clearly in the present case the temperature of the skin surface is lower than that measured. Not being able to determine the true temperature of the skin on account of its small thickness, we are forced to credit it with the bath temperature, which is actually higher.

Study of the radiating power of a steel bath containing added chromium fully confirms the results obtained with vanadium.

Within the limits of experimental accuracy, no explicit temperature dependence of radiating power could be found for iron and steel baths in the near infrared part of the spectrum (roughly from 0.8 to 2.0 μ) over the temperature range studied.

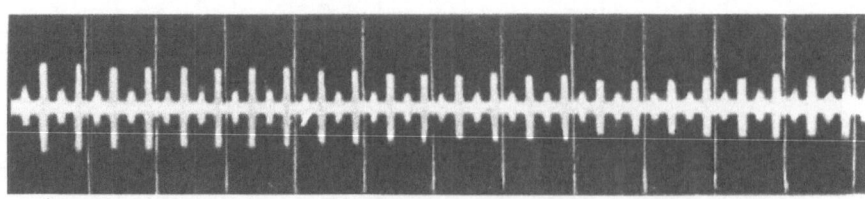

Fig. 123. Reflection picture of the dynamics of growth for an oxide film on molten cast iron.

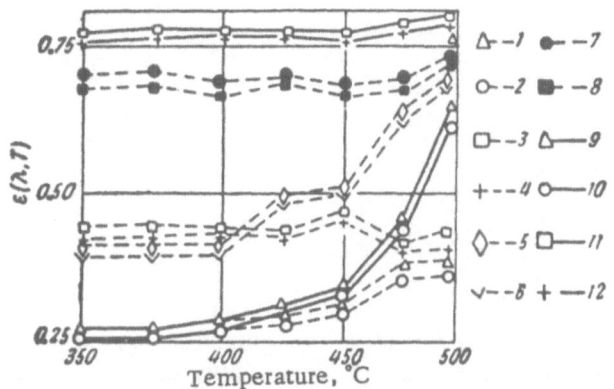

Fig. 124. Temperature dependence of $\varepsilon(\lambda_1, T)$, $\varepsilon(\lambda_2, T)$
for alloy D-16.

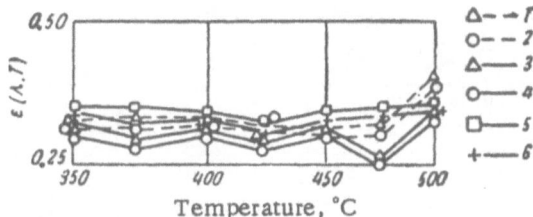

Fig. 125. Temperature dependence of $\varepsilon(\lambda_1, T)$,
$\varepsilon(\lambda_2, T)$ for alloy V-65.

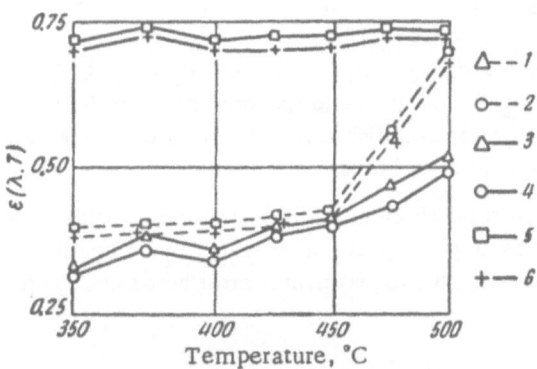

Fig. 126. Temperature dependence of $\varepsilon(\lambda_1, T)$,
$\varepsilon(\lambda_2, T)$ for alloy AK-4.

Since such a dependence exists for the visible region, it may be supposed that the reason for the practically imperceptible temperature dependence is that the measurements were made in the region of almost zero temperature coefficient of the radiating power of iron [22].

The change in spectral reflecting power during the growth of an oxide film on the surface of molten cast iron is seen very clearly in Fig. 123, taken from the paper by Samarin and Svet [120]. Here we see an oscillogram of pulses of modulated radiation in the red $\lambda = 0.62 \mu$ ("larger" pulses) and blue $\lambda = 0.46 \mu$ ("smaller" pulses) parts of the spectrum, reflected from the surface of a molten cast-iron bath. The oscillogram was taken

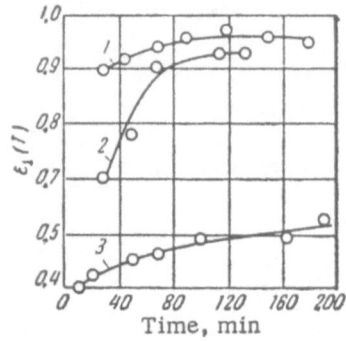

Fig. 127. Variation of the ε_Σ (T) of a ground 4Kh8V2-steel surface with heating time (in min). 1) At 950°C; 2) at 900°C; 3) at 800°C.

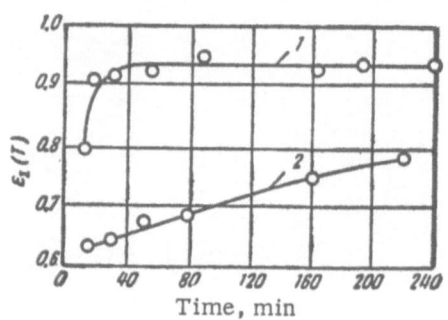

Fig. 128. Variation of the ε_Σ (T) of a ground ÉI481-steel surface with heating time (in min). 1) At 900°C; 2) at 800°C.

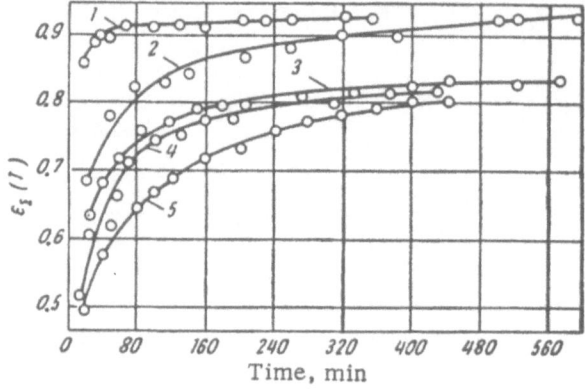

Fig. 129. Variation of the ε_Σ (T) for a ground copper M2 surface with time (in min). 1) At 800°C; 2) at 750°C; 3) at 700°C; 4) at 650°C; 5) at 620°C.

in a modulation reflectometer with rotating light filters. Especially clear is the fall in the two reflected pulses as a result of surface oxidation (film growth). It should be noted that, on the basis of a large number of different experiments, a "gray" character was established for the radiation of the oxide films on the surface of a steel bath containing chromium, vanadium, and other components.

The spectral radiating power of solid aluminum alloys for surfaces with various kinds of finish was determined in [121] for two wavelengths ($\lambda \approx 1.6 \mu$ and $\lambda \approx 1.9 \mu$) by comparing the radiation of the surface in question with that of a black-body cavity, the experimental error being 5%. The results appear in Figs. 124-126.

Alloy D-16 (Fig. 124)

Curves 1 and 2 show the changes in $\varepsilon(\lambda_1, T)$ and $\varepsilon(\lambda_2, T)$ for a strictly clean surface. Mechanical treatment was effected without emulsion cooling. In the temperature range 350 to 425°C, $\varepsilon(\lambda, T)$ varies little. Thereafter $\varepsilon(\lambda, T)$ increases. The increase is especially marked at 450°C. This rise in $\varepsilon(\lambda, T)$ is associated with the appearance of an oxide film on the surface of the metal. On second heating $\varepsilon(\lambda, T)$ alters very little, since a stable oxide film has been formed on the surface (curves 3 and 4).

For a coarse finish (curves 5 and 6), the form of variation of $\varepsilon(\lambda, T)$ remains as before, although the absolute values increase. On repeated heating (7, 8), $\varepsilon(\lambda, T)$ changes very little. Analogous curves were also obtained for a polished surface (curves 9, 10, 11, 12).

The value of $\varepsilon(\lambda, T)$ increases as before, and at a constant temperature of 500°C is independent of time, repeated heating producing no change (curves 11 and 12).

Alloy V-65 (Fig. 125)

This alloy has good anticorrosion properties at temperatures 350 to 500°C. As we may judge from curves 1 and 2 (first heating, surface coarsely finished), 3 and 4 (second heating), and 5 and 6 (polished surface), $\varepsilon(\lambda, T)$ varies very little.

Alloy AK-4 (Fig. 126)

The manner in which the $\varepsilon(\lambda, T)$ of alloy AK-4 varies is analogous to that of D-16. The graphs contain: curves 1 and 2 for a coarsely finished surface (first heating), 3 and 4 for a polished surface (first heating), and 5 and 6 for a polished surface (second heating). At temperatures of 450 to 500°C, $\varepsilon(\lambda, T)$ rises sharply. On repeated heating, $\varepsilon(\lambda, T)$ changes hardly at all.

A large treatise by Rudnaya and Bostrem [122] is devoted to the study of total radiating power for various metal surfaces in the course of oxidation; data describing the variation of $\varepsilon_{\Sigma}(T)$ for ground surfaces with heating time, taken from this, appear in Figs. 127-129.

Appendix I

Spectral Radiating Power $\varepsilon(\lambda, T)$ for the Oxides of Metals and Oxidized Alloys
for Wavelength $\lambda = 0.65\ \mu$ [61] and [123].

Oxides of metals and oxidized alloys	Range of measured values	Value of oxidized polished surface of metals and alloys
Aluminum	0.22—0.40	0.30
Beryllium	0.07—0.37	0.35
Vanadium	—	0.70
Iron	0.63—0.96	0.70
Yttrium	—	0.60
Cobalt	—	0.75
Magnesium	0.1—0.43	0.20
Copper	0.6—0.8	0.70
Nickel	0.85—0.96	0.90
Niobium	0.55—0.71	0.70
Tin	0.32—0.60	—
Uranium	—	0.3
Titanium	—	0.5
Chromium	0.6—0.8	0.7
Cerium	0.58—0.8	—
Zirconium	0.18—0.43	0.4
Alumel	—	0.87
Chromel		0.70
90,10	—	0.87
80,20	—	0.9
60,24,16	—	0.83
Constantan	—	0.84
Carbon steel	—	0.80
Stainless steel	—	0.80

Appendix II

Total Radiating Power of Unoxidized Metals

Metals and Alloys	State solid	State liquid	Metals and Alloys	State solid	State liquid
Beryllium	0.61	0.61	Carbon	0.80-	—
Vanadium	0.35	0.32		0.93	
Tungsten	0.43	—	Uranium	0.54	0.34
Iron	0.35	0.37	Chromium	0.34	0.39
Gold	0.14	0.22	Erbium	0.55	0.38
Yttrium	0.35	0.35	Zirconium	0.32	0.30
Iridium	0.30	—	Iron	0.37	0.40
Cobalt	0.36	0.37	Constantan	0.35	—
Columbium	0.37	0.40	Monel	0.37	—
Manganese	0.59	0.59	Steel	0.35	—
Copper	0.10	0.15	Chromel P:		
Molybdenum	0.37	0.40	90% Ni, 10% Cr	0.35	—
Nickel	0.36	0.37	80% Ni, 20% Cr	0.35	—
Palladium	0.32	0.37	60% Ni, 24% Fe,		
Platinum	0.30	0.38	16% Cr	0.36	—
Rhodium	0.24	0.30	Alumel:		
Silver	0.07	0.07	95% Ni, Al, Mn	0.37	—
Tantalum	0.49	—	Stainless steel:		
Titanium	0.63	0.65	27% Cr, 18% C,		
Thorium	0.36	0.40	10% T	—	0.64

LITERATURE CITED

1. G. S. Landsberg. Optics, GTTI (1947).
2. G. W. Strett (Lord Rayleigh). Theory of Sound, Vol. 1 [Russian translation; S. M. Rytov, editor], GTTI (1955), p. 398.
3. L. I. Mandel'shtam. Collected Works, Vol. 1, Akad. Nauk SSSR (1948), p. 246.
4. L. M. Brekhovskikh. Zh. Éksperim. i Teor. Fiz., No. 3(9): 289 (1952).
5. M. A. Isakovich. Zh. Éksperim. i Teor. Fiz., No. 3(9): 305 (1952).
6. G. Mie. Ann. Phys. 25: 377 (1908).
7. H. O. McMahon. J. Opt. Soc. Am. 40(6): 376 (1950).
8. W. Finkelburg. J. Opt. Soc. Am. 39: 185 (1949).
9. G. Ribault. Optical Pyrometry [Russian translation from the French; G. S. Landsberg, editor], GTTI (1934).
10. G. Gräber, S. Erk, and U. Grigul. Fundamentals of the Science of Heat Exchange [Russian translation from the German], IL (1952).
11. M. Born. Optics [Russian translation], DNTVU (1937).
12. Temperature and Its Measurement, Vol. II. Collection of articles [Russian translation], IL (1954).
13. W. F. Forsythe, Worthing. Astrophys. J. 61: 165 (1925).
14. Worthing, J. Opt. Soc. Am. 13: 635 (1926).
15. M. Czerny. Z. Physik 26: 215 (1924).
16. H. Bethe and A. Sommerfeld. Electron Theory of Metals [Russian translation from the German], ONTI (1938).
17. Krönig. Proc. Roy. Soc., Vol. 124 (1929); Vol. 133 (1931).
18. A. V. Sokolov. Optical Properties of Metals, Fizmatgiz (1961).
19. J. A. Stratton. Theory of Electromagnetism [Russian translation], Gostekhizdat (1948).
20. O. Lummer and E. Kurlbaum. Verhandlungen der Deutschen Physikalischen Gesellschaft 17: 106—111 (1898); 1: 215-235 (1899).
21. N. W. Snyder. Trans. ASME, May: 541 (1954).
22. D. J. Price. Proc. Phys. Soc. 59 (331): 1 (1947).
23. L. Ward. Proc. Phys. Soc. 70 (380): 343 (1956).
24. D. Ya. Svet and O. N. Talenskii. Transactions of the A. A. Baikov Institute of Metallurgy, Akad. Nauk SSSR, No. 5: 183 (1960).
25. C. P. Butler and C. Y. Edwards. AEMM 1: 2 (1957).
26. J. T. Bevans, J. T. Gier, and R. V. Dunkle. Trans. ASME, Paper 57-A-29: 1—10 (January 10, 1958).
27. G. Liebmann. Z. Physik 63: 404 (1930); 71(78): 416 (1931).
28. M. Michaud. Silicates Ind. 19(617): 243 (1954).
29. R. Allegre. Ann. Phys. (Paris) (1954); Ceramic Abstr. (November 1955).
30. B. S. Kellett. J. Opt. Soc. Am. 42(5): 339 (1952); J. Soc. Glass Technol. 36: 115—231 (1952).
31. M. Czerny and L. Genzel. Glasstech. Ber. 25(5): 134—139 (1952).
32. R. Gardon. J. Am. Ceram. Soc. 39(8): 278 (1956).
33. R. Gardon. J. Am. Ceram. Soc. 41(6): 200 (1958).
34. J. R. Beattie and E. Coen. Brit. J. Appl. Phys. 11(4): 151 (1960).
35. B. V. Stark and Yu. M. Shashkov. Radiating Power of Metals, Vol. 1, Izv. Akad. Nauk SSSR, Otd. Tekhn. Nauk 1(3) (1952).
36. P. M. Reynolds. Brit. J. Appl. Phys. 12: 111 (March 1961).
37. R. Hase. Z. Tekh. Physik 13(13): 145 (1932).
38. D. Ya. Svet. Tr. VNITOM 5: 100 (1953).
39. D. Ya. Svet, V. V. Grishin, and S. L. Naryshkin. Modulation Reflectometer, VINITI (1958).

40. J. T. Gier, R. V. Dunkel, and J. T. Bevans. J. Opt. Soc. Am. 44(7): 558 (1954).

41. D. Ya. Svet. Dokl. Akad. Nauk SSSR 129(6) (1959).

42. V. A. Osipova. Teploénerg., p. 59 (1960).

43. A. Vinokurov. Izv. V. I. Ul'yanov-Lenin LÉTI 35: 20 (1958).

44. D. M. Bowie. Nat. Couv. Rec. MTT 15(1) (1957).

45. De Vos. Physica, No. 20: 669 (1954).

46. R. D. Larrabee. J. Opt. Soc. Am. 49(6): 619 (1959).

47. Secretary's Report on the Tenth Conference of the International Illuminescence Commission, 1939, Vol. I (Scheveningin, 1942).

48. L. S. Ornstein. Physica 3: 561 (1936).

49. T. Riethof, B. D. Acchione, and E. R. Branyan. Temperature: Its Measurement and Control in Science and Industry, Vol. III, Part 2, Appl. Methods and Instruments (Reinhold Publishing Corp., New York), p 515.

50. K. L. Kudkin, W. I. Parker, and K. J. Jenkins. Temperature: Its Measurement and Control (see ref. 49).

51. W. F. Forsythe. Smithsonian. Phys. Tables, 9th edition (Washington, D. C. 1954).

52. R. Spulle. Z. Physik 72: 25 (1931).

53. J. Worthing. Phys. Rev. 25: 846 (1925); 28: 135 (1926); 28: 174−200 (1926).

54. J. T. Gier, R. V. Dunkle, J. T. Bevans. J. Opt. Soc. Am. 44 (7) (1954).

55. O. Levi and Espersen. Metallwirtschaft, p. 1063 (1930); Phys. Rev. 78: 231 (1950); J. Metals, No. 1, Sect. 2: 168 (1955).

56. Sims, Graighead, and Jaffe. Trans. Am. Inst. Mining Met. Eng., No. 203: 168 (1955).

57. D. Marple. J. Opt. Soc. Am. 46 (7): 490 (1956).

58. D. Marple. J. Opt. Soc. Am. 48(9): 300 (1956).

59. O. Baird. Platinum Metals Rev. 4(1): 31 (1960).

60. R. W. Douglass and E. F. Adkins. Trans. Metallurg. Soc. AIME 221: 30 (April 1961).

61. J. Worthing. Temperature. Its Measurement and Control in Science and Industry, Vol. II (Reinhold Publishing Corp., New York), [translation of article in the collection "Methods of Temperature Measurement," IL (1954)].

62. J. Spence. Astrophys. J., No. 37: 194 (1913).

63. J. Worthing. Phys. Rev., No. 28: 174 (1926).

64. J. Stephens. J. Opt. Soc. Am. 29: 158 (1939).

65. R. Börnstein and H. Landolt. Phys. Chem., Table 3, 1: 3 (1935).

66. A. M. Samarin and D. Ya. Svet. Experimental Techniques and Methods of Investigations of High Temperatures, Akad. Nauk SSSR (1959).

67. M. N. Dastur and N. A. Gosken. J. Metals, No. 11: 188 (1950).

68. L. Ward. Proc. Phys. Soc., Ser. B. 69(435): 339 (1956).

69. L. S. Ornstein and Vander Veen. Phys. 6: 439 (1939).

70. R. Hase. Z. Physik 15: 54 (1923).

71. J. Coblentz. Publication No. 65 (Carnegie Institute, Washington, D. C. 1906).

72. D. J. Price and H. Lowery. J. Iron Steel Inst. (London) 491(1): 523 (1944). Paper No. 7/1943.

73. W. Pepperhoff. Z. Angewandte Phys. 12(4): 168 (1960).

74. C. M. Stubbs. Proc. Roy. Soc. 88: 195 (1913).

75. Yu. A. Vykhovskii. Tsvetnye Metally, No. 4: 27 (1953).

76. C. M. Stubbs and H. Prideaux. Proc. Roy. Soc. 87: 451 (1912).

77. R. Wood. Physical Optics [Russian translation], ONTI (1936).

78. C. M. Stubbs. Proc. Roy. Soc., Vol. 88 (1913).

79. D. Smith and F. Chapmen. Trans. Am. Inst. Min. Metal, Eng. J. Metals 4: 643 (1952).

80. A. M. Samarin and D. Ya. Svet. Dokl. Akad. Nauk SSSR 126(1) (1959).

81. D. Ya. Svet. Heat-Resistant Alloys, Vol. II, Akad. Nauk SSSR (1960), p. 323.

82. J. Poggie, Prod. Eng., p. 205 (September 1953).

83. J. T. Bevans, J. T. Gier, and R. V. Dunkle. Trans. ASME 10:1 (1958). Paper No. 57-A-29.

84. A. G. Bloch. Fundamentals of Heat Exchange by Radiation [Russian translation], Gosénergoizdat (1962).

85. F. C. Allen. J. Appl. Phys. 28(12) (1957).

86. V. S. Vavilov, A. G. Ginius, and M. M. Gorshkov. Zh. Tekhn. Fiz. 28(2): 254 (1958).

87. T. S. Moss and T. D. Hawkins. Proc. Phys. Soc. 72: 270 (1958).

88. J. D. Plankett and W. D. Kingery. Proceedings of the Second Conference on Carbon (Oxford, 1960), p. 457.

89. J. Euler. Ann. Phys. Lpz. 11: 203 (1953).

90. H. G. MacPherson. J. Opt. Soc. Am. 30: 189 (1940).

91. F. Patzelt and Baldewein. Wiss. Veröff. Siemens-Konzern 21: 213 (1943).

92. W. W. Coblentz. J. Res. Natl. Bur. Std. Bull. 7: 197 (1911).

93. S. P. F. Humphrey-Owen and L. A. Gilbert. Industrial Carbon and Graphite (Society of Chemical Industry, London, 1958), p. 37. (Distributed by Macmillan.)

94. H. T. Betz, O. H. Olson, B. D. Schurin, and J. C. Morris. W. A. D. Tech. Rept., Part II, pp. 56-222 (1957).

95. R. J. Thorn and O. C. Simpson. J. Appl. Phys. 24: 633 (1953).

96. S. C. Jain and K. S. Krishnan. Proc. Roy. Soc., No. 225 (1954).

97. T. Royd. Phil. Mag., No. 21: 167 (1911).

98. W. W. Coblentz. J. Res. Natl. Bur. Std. Bull., No. 9: 283 (1913).

99. L. Harris and K. F. Cuff. J. Opt. Soc. Am., No. 46: 160 (1954).

100. A. H. Pfund. Measurement of Radiant Energy, W. E. Forsythe, editor (New York, McGraw-Hill Book Company, 1937, pp. 210-215.

101. E. K. Plyler and J. J. Ball. J. Opt. Soc. Am., No. 38: 988 (1948); R. H. Heilman. Trans. Am. Soc. Mech. Engrs., No. 51: 287 (1929); B. T. Barnes. Phys. Rev., No. 34: 1026 (1929).

102. A. H. Sully, E. A. Brandes, and R. B. Waterhouse. Brit. J. Appl. Phys., No. 3: 97 (1952).

103. A. Umur, G. V. Parmelle, and E. Schutrum. Heating, Piping Air Conditioning, No. 26: 135 (1954).

104. C. Krijgsman. Physica, No. 5: 918 (1938).

105. H. G. MacPherson. American Institute of Physics (Reinhold Publishing Corp., New York, 1941), pp. 1141-1149.

106. D. M. Parker and C. Lock. J. Opt. Soc. Am., No. 42: 879 (1952).

107. F. S. Johnson. J. Opt. Soc. Am., No. 46: 101 (1956).

108. M. R. Null and W. W. Lozier. J. Appl. Phys., No. 29: 1605 (1958).

109. K. Warmuth. Wiss. Veröff. Siemens 7: 307 (1928); J. T. McCartney and S. Ergun. Fuel 37: 272 (1958).

110. F. Anacker and R. Mannkopff. Z. Physik 155: 16 (1959).

111. S. Ergun and J. T. McCartney. Fuel 34: 71 (1960).

112. R. Allegre. Silicates Ind. 23(5) (1958).

113. G. R. Blair. J. Am. Ceram. Soc. 43(4): 197 (1960).

114. A. Sokolov, B. Gorban', et al. Opt. i Spektroskopiya 2(12): 772 (1961).

115. H. H. Blau. Proceedings of International Symposium of High Temperature Technology (McGraw-Hill Book Company, Inc., New York, 1959), p. 45. [Russian translation in the book Studies at High Temperatures, Academician V. A. Kirilin, editor, INO (1962)].

116. F. Gaus. Revue d'Optique 10: 491 (1954).

117. G. V. Samsonov and I. I. Pen'kovskii. Opt. i Spektroskopiya, Akad. Nauk SSSR, Vol. 11, No. 3.

118. A. A. Sterlyadkina. Radiation of Glass, VNIIStekla (1956).

119. D. Ya. Svet. Problems of Metallurgy, Akad. Nauk SSSR (1959), p. 99.

120. A. M. Samarin and D. Ya. Svet. Dokl. Akad. Nauk SSSR 108 (1): 79 (1956).

121. A. A. Poskachei and D. Ya. Svet. Metallurgiya Topliva, No. 3: 86 (1960).

122. A. I. Rudnaya and Z. D. Bostrem. Studies in the Field of Thermal Measurements, Trudy VNIIM, No. 35(95): 95 (1958).

123. T. R. Harrison. Radiation Pyrometry and Its Underlying Principles of Radiant Heat Transfer (J. Wiley and Sons, Inc., New York, 1960) [Russian translation, D. Ya. Svet, editor, Mir (1964)].